William Fogg OSGOOD
AT HARVARD

Agent of a Transformation
of Mathematics in the United States

DIANN R. PORTER

Docent Press
Boston, Massachusetts, USA
www.docentpress.com

Docent Press publishes books in the history of mathematics and computing about interesting people and intriguing ideas. The histories are told at many levels of detail and depth that can be explored at leisure by the general reader.

Cover design by Brenda Riddell, Graphic Details.

Produced with TEX. Textbody set in Garamond with titles and captions in Bernhard Modern.

The picture of William Fogg Osgood on the cover and on page 5 by permission of Harvard University Archives, HUP Osgood, W.F. (1). The picture of the portrait of Nathaniel Bowditch by Chester Harding on page 38 by permission of the American Antiquarian Society. The picture of the presidents of the American Mathematical Society on page 152 by permission of the American Mathematical Society.

© Diann R. Porter 1997, 2013

All rights reserved. No part of this book may be reproduced or utilized in any form or by any means, electronic or mechanical, including photocopying and recording, or by any information storage and retrieval system, without permission in writing from the author.

Acknowledgements

At the beginning of my career, a number of years ago, two people helped set me on a positive path. I take this opportunity to offer thanks.

I am deeply indebted to Professor Emeritus William A. Howard of the University of Illinois at Chicago. I am thankful for all the time he invested in teaching me mathematics and discussing ideas with me. His support, guidance and enthusiasm made working on this research an enjoyable and enriching experience. Throughout my career I have benefitted from habits of thought and work I learned by working with him—and for this I am truly grateful.

I am also grateful to Karen Hunger Parshall for the idea that eventually developed into this research. I thank her for always having found gracious ways to offer constructive criticism. Most of all, I am thankful for her encouragement and for inspiring me to an interest in the mathematical community in the United States around the turn of the century.

Contents

1 Transformation in the Air 1
   1.1 The Lives of William Fogg Osgood and Maxime Bôcher 3
   1.2 Historical Periods Spanned by Osgood's Career 10
   1.3 The New Complete Mathematics Professional 17
   1.4 Membership in the New Community of Mathematicians 17
   1.5 A Responsible Educator in a New Style 21
   1.6 Top Mathematical Researcher 23
   1.7 Osgood's Harvard 24
   1.8 Osgood's Research and Teaching—A Taste of Chapters to Come . 26

2 Before Osgood and Bôcher—Signs of a Transformation to Come 29
   2.1 Building a Mathematical Foundation: 1727–1869 31
   2.2 Charles Eliot's Vision 53
   2.3 Peirce to Osgood and Bôcher 59

3 Non-Uniform Convergence and the Integration of Series Term by Term 65
   3.1 Some Historical Context 68
   3.2 Non-Uniform Convergence 75
   3.3 Part II of the 1887 Paper—Term by Term Integration 80
   3.4 Reaction to Osgood's Paper and its Consequences 87

4 Osgood's Proof of the Riemann Mapping Theorem 93
   4.1 Setting and Structure of Osgood's Paper 94
   4.2 Historical Context—It was "in the air" 98
   4.3 Osgood vs. Poincaré and Harnack 103
   4.4 Osgood's Proof 110
   4.5 Reactions and Consequences 118

5 A Jordan Curve of Positive Area 123
   5.1 Motivation for Osgood's Result 124
   5.2 Historical Context—What is *area*? 125
   5.3 Osgood's Construction of the Curve 129
   5.4 Enduring Interest in Osgood's Result 137

6	Osgood's Teaching—Textbooks, Students and Approaches	141
	6.1 Textbooks, Calculus Reform and the Use of Infinitesimals	144
	6.2 Lehrbuch der Funktionentheorie	158
	6.3 Bôcher's Introduction to Higher Algebra	167
	6.4 The Students of Osgood and Bôcher	171
A	Letters of Angus E. Taylor to Diann Porter	201
	A.1 January 25, 1997	202
	A.2 January 25, 1997	203
	A.3 February 15 and 16, 1997	204
	A.4 February 16, 1997	211
	A.5 May 9, 1997	212
	A.6 June 6, 1997	213
	A.7 September 12, 1997	215

List of Figures

1.1	William Fogg Osgood	5
1.2	Maxime Bôcher	6
1.3	Frank Nelson Cole	19
1.4	E.H. Moore	21
2.1	Nathaniel Bowditch by Chester Harding	38
2.2	Benjamin Peirce	41
3.1	Title Page of Lebesgue's Thesis	72
5.1	Osgood's curve construction—stage one	131
5.2	Osgood's curve construction—stage two	132
6.1	Presidents of the American Mathematical Society	152
6.2	G. D. Birkhoff	176

Chapter 1

Transformation in the Air

In the turn from the 19th to the 20th century, reform in education at all levels was in the air in the United States. Particularly in mathematics, a strong movement formed toward initiating graduate education and then modernizing undergraduate education.

American students of many fields in the mid-19th century experienced the "lure of the German University."[1] In the first half of the century, some 200 Americans studied in German universities. In the second half, the floodgates opened resulting in roughly 9,000 students pursuing studies there, about 2,000 of them in the 1880s alone.[2] The numerical peak for Americans in Germany occurred in 1895–1896 with 517 Americans officially matriculated[3], but by that time returning students had already begun to instigate major changes in the American mathematical community. The American students enjoyed the atmosphere created in the German universities by three ideals: *Wissenschaft*, which represented the search for pure scientific knowledge through orderly inquiry, *Lehrfreiheit*, or freedom of teaching, and *Lernfreiheit*, or freedom of learning. Fe-

---
[1] See Veysey, 1965, p. 124.
[2] See Cremin, 1988, p. 557.
[3] See Veysey, 1965, p. 130.

lix Klein of Göttingen acted as a magnet for American students of mathematics.

The American students returned from Germany determined to share with institutions in their own country their excitement for German-style pure research, teaching and learning. Their interest found resonance and sustenance in men like Johns Hopkins, Ezra Cornell, Jonas Clark, and John D. Rockefeller who contributed their money to help establish the Johns Hopkins, Cornell and Clark Universities and the University of Chicago as institutions that incorporated some of this German style. In the 1880s and 1890s Morrill land grants and other government funding started to become available for the effort. Harvard's president, Charles William Eliot found money among wealthy Bostonians to transform his institution, partly to compete with the Johns Hopkins and partly to realize his Americanized vision of the German model.[4]

William Fogg Osgood and Maxime Bôcher were two of the Americans returning from Germany into this atmosphere of reform. They both received their Ph.D.s in mathematics in Germany, under the influence of Felix Klein of Göttingen. Both Osgood and Bôcher returned from Germany to join the faculty at Harvard, in 1890 and 1891 respectively, where their enthusiasm and German training joined with Eliot's plans for the new American university to produce a transformation in mathematics at Harvard. Osgood and Bôcher were members of the first generation of Americans whose careers are recognizable as those of modern mathematicians, both in research and in work as teachers and mentors to a generation of students.

In particular, Osgood was among the country's top mathematical researchers for a time. He helped put American mathematics on the map of the mathematical world, attracting the attention of prominent European mathematicians such as Henri Lebesgue, Henri

---

[4]See Cremin, 1988, pp. 557–558.

Poincaré, and of course Felix Klein. Osgood enjoyed a long career, beginning during the early years of the American mathematical research community as it manifested itself at Harvard, continuing until his retirement from that institution in 1933 at a time when it possessed a well established world reputation in research mathematics, and ending with two years spent at the National University of Peking.

Although chapters to come focus mostly on Osgood's work, a thread of Maxime Bôcher's work is woven into this historical account because of the way his career paralleled and complemented Osgood's both at Harvard and in the larger American mathematical community. Bôcher's contributions to that community were cut short by his premature death in 1918.

1.1 The Lives of William Fogg Osgood and Maxime Bôcher

William Fogg Osgood was born on 10 March 1864 in Boston, Massachusetts to William and Mary Rogers (Gannett) Osgood. John Osgood of Hampshire, England was one of his ancestors. This early Osgood arrived in Massachusetts in 1638. Mary Rogers Osgood was also of English descent. William Fogg Osgood graduated from the Boston Latin School in 1882, and received his A.B. from Harvard in 1886, graduating *summa cum laude* and ranked second in a class of 286. Much of his early time at Harvard was spent in study of the classics. In mathematics, he enjoyed the influences of mathematical physicist Benjamin Osgood Peirce and Frank Nelson Cole, who already had studied with Felix Klein in Germany. A year of graduate work at Harvard resulted in a master's degree in 1887. Like others of his generation, he felt the attraction of German higher education, spending 1887–1889 as first a Harris and then a Parker Fellow from Harvard at Göttingen and 1889–1890 at Erlangen, studying with Max Noether in particular. He received his Ph.D. from Erlangen in 1890—his dissertation treating Abelian integrals, a topic in the theory of functions, was based on work of Felix Klein and Max Noether. Osgood married Therese Ruprecht of Göttingen and returned to Harvard to join the faculty. He remained a member of the faculty for 43 years. Osgood's German experiences deeply entwined themselves into his life. In addition to having a German wife, he wrote his major works on the theory of functions in German and was said to take on German mannerisms. It has also been reported that he sympathized with Germany in World War I. Osgood and his first wife divorced, and he married Celeste Phelps Morse (formerly wife of Osgood's colleague Marston Morse) in 1932. After his retirement from Harvard in 1933, he taught for two years, 1934–1936, at the National University of Peking, as it was then called. Historian Raymond Clare Archibald noted that Osgood's favorite pastime was "touring in his motor car." He died on 22 July 1943 in Belmont, Massachusetts.[5]

Aspects of Maxime Bôcher's life and schooling closely resembled those of Osgood. Bôcher was born on 28 August 1867 to Ferdinand and Caroline (Little) Bôcher in Boston, Massachusetts. His paternal grandfather, Ferdinand Jules Bôcher came to America frequently on business from Caen, France. The younger Ferdinand was born in New York and became a well-known professor of modern languages at Harvard. Caroline Little was a member of an old New England family. One of her ancestors was an early member of the Plymouth colony, Thomas Little, who married Anne Warren, the daughter of Richard Warren who came on the Mayflower expedition. Bôcher graduated from the Cambridge Latin School and entered Harvard in 1883 at age 16. He received his A.B. in 1888, graduating *summa cum laude*. Like Osgood before him, Bôcher

---

[5]See Archibald, 1938, p. 153. These and other details of Osgood's life and work can be found in a tribute to Osgood written by his colleague J.L. Walsh. See Walsh, 1989, in his entry in the *Dictionary of Scientific Biography* (v. 5, 1974), which was also written by Walsh, and in R.C. Archibald's *A Semicentennial History of the American Mathematical Society 1888–1938*. See also Archibald, 1938, which contains a bibliography of Osgood's work.

1. Transformation in the Air

Figure 1.1: William Fogg Osgood

was both a Harris and a Parker Fellow from Harvard at Göttingen, where he studied with Felix Klein and others from 1888 to 1891, receiving his Ph.D. in 1891. He returned to serve on the faculty at Harvard until his early death in 1918.[6]

There were striking parallels in the careers of the two men. Osgood graduated from Harvard in 1886, and Bôcher graduated in 1888. (This was also the year the American Mathematical Society was formed in New York.) Both men received scholarships for further study in Germany, and both studied with Felix Klein. Both wrote articles for the *Enzyklopädie der Mathematischen Wissenschaften* (Encyclopaedia of Mathematical Sciences) at Klein's

---

[6]Information about Bôcher's life and work can be found in a memorial written by Osgood, *The Life and Services of Maxime Bôcher*, in Bôcher's entry in the *Dictionary of Scientific Biography* (v. 2, 1974), and in Archibald, 1938. G.D. Birkhoff wrote a detailed summary of Bôcher's research, *The Scientific Work of Maxime Bôcher*. See Birkhoff, 1919.

Figure 1.2: Maxime Bôcher

invitation. Both served as president of the American Mathematical Society, Osgood in 1905–1906 and Bôcher in 1909–1910. When asked in 1903 to rank their top 80 colleagues by the quality of their research, mathematicians gave Osgood a ranking of 3 and Bôcher a ranking of 4. (E.H. Moore and G.W. Hill were ranked first and second, respectively.)[7]

Both men spent the major part of their mathematical careers at Harvard. Upon returning from Germany, Osgood served as instructor 1890–1893, assistant professor 1893–1903, and professor 1903–1933. Closely following this path, Bôcher served as instructor 1891–1894, assistant professor 1894–1904 and professor until 1918, the year of his death. Bôcher, along with James Pierpont of Yale, gave the first American Mathematical Society Colloquium Lectures in 1896. The mathematical community was still very small—13

---
[7]See Archibald, 1938, pp. 153, 161.

people attended the lectures. Osgood, along with A.G. Webster, gave the second series of Colloquium Lectures in 1898.

Similarities aside, Garrett Birkhoff noted major differences in their mathematical personalities. He called Osgood "rigorous, systematic, and thorough." His theorems in analysis were "notable for their sharpness and Weierstrassian rigor." In contrast, Bôcher was "intuitive, brilliant, and fluent," and his "monographs on integral equations and on the methods of Sturm were models of lucidity and perception, as was his invited address at the 1912 International Mathematical Congress."[8] R.C. Archibald wrote: "All of Bôcher's papers excel in simplicity and elegance and nearly all of them treat subjects of great importance to marked advantage. He never occupied himself with an unimportant problem."[9]

The names of the two men, Osgood and Bôcher, are often linked in historical accounts of the mathematics department at Harvard. They formed a "Harvard Team" in many respects. This book, however, focuses on Osgood, the senior member of the team whose career path and striking research results provide a leading example of the turn-of-the-century transformation of the American mathematical arena.

---

[8] See Birkhoff, 1977, pp. 33–34.
[9] See Archibald, 1938, pp. 162–163.

William Fogg Osgood

1864	Born 10 March in Boston, Massachusetts.
1882	Graduated from Boston Latin School. Began undergraduate work at Harvard.
1886	A.B. from Harvard *summa cum laude*.
1887	A.M. from Harvard. Began studying at Göttingen with Felix Klein.
1889	Transferred to Erlangen.
1890	Received his Ph.D. from Erlangen. Thesis title: *Zur Theorie der zum algebraischen Gebilde $y^m = R(x)$ gehörigen Ableschen Functionen* (On the theory of algebraic structures of Abelian functions $y^m = R(x)$. Married Therese Ruprecht of Göttingen. Began teaching at Harvard.
1893	Promoted to assistant professor.
1897	Published a major result on non-uniform convergence and term by term integration.
1898	Gave AMS Colloquium Lectures on topics in the theory of functions.
1900	Published his proof of the Riemann mapping theorem.
1901	In the period 1901–1913, directed his only four doctoral students: C.W. McG. Black, L.D. Ames, E.H. Taylor and G.R. Clements (with C.L. Bouton).
1903	Promoted to full professor. Vice President, AMS. Exhibited a Jordan curve of positive area.
1904	Member, National Academy of Sciences.
1905	President, AMS.
1907	Published the treatise *Lehrbuch der Funktionentheorie* and textbook *A First Course in the Differential and Integral Calculus*.
1910	Editor, *Transactions of the AMS*.
1913	AMS Colloquium Lectures—for the second time.
1922	Acting Dean, Graduate School of Arts and Sciences, Harvard (February–July).
1932	Married Celeste Phelps Morse.
1933	Retired from Harvard.
1934	Taught at National University of Peking (until 1936).
1943	Died 22 July. Buried at Forest Hills Cemetery in Boston.

Table 1.1: Some Biographical Details of William Fogg Osgood

1. Transformation in the Air

Maxime Bôcher

1867	Born 28 August in Boston, Massachusetts.
1883	Graduated at Cambridge Latin School. Entered Harvard.
1888	A.B. Harvard *summa cum laude*. Began studies at Göttingen (Harvard, Harris and Parker Fellow) where he studied with Klein, Schoenflies, Schur, Schwarz and Voigt.
1891	Ph.D. from Göttingen. Joined the faculty at Harvard.
1894	Promoted to assistant professor at Harvard. Published the book *Ueber die Reihenentwickelungen der Potentialtheorie* (Development of the potential function into series), an elaboration of his dissertation.
1895	Beginning in 1895, directed the dissertations of 17 students: J.W. Glover, M.B. Porter, F.H. Safford, D.F. Campbell, O. Dunkel, D.R. Curtiss, W.B. Ford, W.H. Reover, W.C. Brenke, F. Irwin, C.N. Moore, G.C. Evans, T. Fort, L.R. Ford, M.T. Hu, L. Brand, C.N. Reynolds (with Birkhoff).
1896	Gave the first AMS Colloquium Lectures jointly with James Pierpont.
1902	Vice-president, AMS.
1904	Promoted to full professor at Harvard.
1907	Published *Introduction to Higher Algebra*. Reprinted 14 times by 1937. Translated into German and Russian.
1908	Editor-in-chief, *Transactions of the AMS*.
1909	President, AMS. Member, National Academy of Sciences.
1912	Lectured on "Boundary problems in one dimension" at the International Congress of Mathematicians at Cambridge, England.
1913	Exchange professor at the University of Paris.
1915	Published the texts *Plane Analytic Geometry with Introductory Chapters on the Differential Calculus*, and *Trigonometry with the Theory and Use of Logarithms*.
1918	Died 12 September in Cambridge, Mass.

Table 1.2: Some Biographical Details of Maxime Bôcher

1.2 Historical Periods Spanned by Osgood's Career

In their book, *The Emergence of the American Mathematical Research Community, 1876–1900: J.J. Sylvester, Felix Klein, and E.H. Moore*, Karen H. Parshall and David E. Rowe set out a periodization of American mathematics. In the first period, 1776–1876, the mathematical endeavor in the new United States was closely tied to the larger scientific effort to find an American rather than a colonial scientific identity. The American scientists measured their progress by the standards of their European models.

During the second period, 1876–1900, an *American mathematical research community* formed. The group of American mathematicians was indeed a *community*, with the American Mathematical Society developing as a cornerstone of its professional structure. This community identified itself as *mathematical*, separating itself from the larger scientific community that was undergoing a generalized process of specialization. It emphasized *research* as a requisite professional activity, both complementary and supplementary to teaching. Furthermore, it was *American*. While the new community looked to European models such as the university in Göttingen for inspiration, the Americans laid the foundations for their own unique professional structure and educational patterns.

A third period, 1900–1933, followed as one of consolidation and growth. The initially small group of high-level American research mathematicians that came out of the emergent period, including Osgood, worked to institutionalize firmly and to strengthen the ideals that shaped American mathematics in the quarter century from 1876 to 1900. The new American mathematician participated in an American Mathematical Society which had grown from a small band into a large national professional organization with a membership composed of researchers who worked in mathematics departments that adopted research as a key professional criterion. Moreover, that mathematician's department had developed the ability to train new research mathematicians which would ensure the future growth of the community. Osgood was a complete mathematics professional as that term came to be understood during the period of consolidation and growth. Delving more deeply into Parshall and Rowe's periodization will provide a rich context for Osgood's contributions.

1.2.1 Period of emergence of a mathematical research community: 1876–1900

Parshall and Rowe identify three men and their institutions as the keys to the emergence of the American mathematical research community—J.J. Sylvester at the Johns Hopkins, Felix Klein at Leipzig and Göttingen, and E.H. Moore at the University of Chicago.

The period began in 1876—the year in which the Johns Hopkins was founded and J.J. Sylvester joined the faculty as its first professor of mathematics. Hopkins was the first research-oriented academic institution in the United States. Sylvester and his graduate students formed a dynamic mathematical research community there, whose work in invariant theory began to catch the attention of European mathematicians. This opening flourish for graduate mathematics education in the United States broke the ice. Unfortunately Sylvester returned to England and the program was no longer able to maintain its position of primacy in training highly competent research mathematicians.[10] Harvard President Charles Eliot had encouraged the innovations at Hopkins, which he looked to, in part, as a model for his institution.

---
[10]For a full account of Sylvester's life and work, see Karen Hunger Parshall's book, *James Joseph Sylvester, Jewish Mathematician in a Victorian World*, published by The Johns Hopkins University Press in May 2006.

With Sylvester's departure, American students of mathematics had to find another mentor and model. These students, drawn by the "lure of the German university," were welcomed by German mathematician Felix Klein. The German universities had developed a system of seminars for training mathematics students. Parshall and Rowe note that: "Perhaps the greatest advantage of the seminar system at the German universities was that it took students to the threshold of research activity much more quickly than would otherwise have been possible."[11] Klein was a skilled teacher whose seminars were highly organized and tightly controlled. His "dedication to and interest in mathematics education ultimately proved to be a characteristic feature within the early American mathematical community."[12]

Klein's first American student was Irving Stringham from Hopkins who arrived at the university in Leipzig in 1880—Klein's first year there. Frank Nelson Cole and Henry Burchard Fine joined Klein's lectures in 1884, but did not attend his seminars. Klein moved to Göttingen in the summer of 1886 and the number of Americans studying with him gradually increased. Mellen Woodman Haskell, who had received his bachelor's and master's degrees at Harvard, arrived that first summer. Cole returned to Germany to study with Klein in 1887, along with Henry Dallas Thompson who came from Princeton. But Klein's star American students at Göttingen were William Fogg Osgood and Maxime Bôcher who journeyed there in 1887 and 1888, respectively. In the years that followed, increasing numbers of American students arrived to participate in Klein's lectures and seminars.[13]

Klein's role as mentor and shaper of the emerging American mathematical research community continued in full force up through his visit to the United States at the time of the 1893 Math-

---

[11] See Parshall and Rowe, 1994, p. 190.
[12] See Parshall and Rowe, 1994, p. 191.
[13] See Parshall and Rowe, 1994, Chapter 5, entitled "America's *Wanderlust* Generation," for details about the studies of these and other American students of Klein at Leipzig and Göttingen.

ematical Congress held in conjunction with the World's Columbian Exposition in Chicago. Around that time, a new figure began to emerge as the dominant shaper of the community—E.H. Moore, Professor of Mathematics at the fledgling University of Chicago.

The University of Chicago opened on 1 October 1892. E.H. Moore had been chosen by President William Rainey Harper to head the mathematics department. Moore and his colleagues, including German mathematicians Oskar Bolza and Heinrich Maschke, put together strong undergraduate and graduate programs. The undergraduate program was designed both to expose undergraduates in general to mathematical culture and to prepare undergraduates in mathematics for graduate study. The graduate program was specifically designed to prepare the student for *serious* mathematical research. Leonard Eugene Dickson became Moore's first doctoral student—he received his Ph.D. in 1896. He was followed by men who would also become the cream of the crop of the second generation of American research mathematicians, Oswald Veblen, Robert L. Moore (of "Moore Method" fame) and George David Birkhoff, for example. The mathematics program at the University of Chicago quickly became the best in the nation.[14]

During the years following their return from Germany, Osgood and Bôcher were Moore's counterparts in the mathematics department at Harvard. While Moore was able to build his department from the ground up, Osgood and Bôcher had to work a gradual transformation of the existing department. G. D. Birkhoff's son, Garrett Birkhoff wrote: "Harvard's national leadership and international reputation during the decades 1894–1914 were primarily due to William Fogg Osgood (1864–1943) and Maxime Bôcher (1867–1918)."[15] He also believed that, due to Felix Klein's influence, Osgood and Bôcher "spearheaded a revolution in mathemat-

---

[14] See Parshall and Rowe, 1994, Chapter 9, entitled "Meeting the Challenge: The University of Chicago and the American Mathematical Research Community" for further details.
[15] See Birkhoff, 1977, p. 33.

ics at Harvard."[16] Julian Lowell Coolidge might have disagreed with Birkhoff's later assessment. In a 1930 history of Harvard, Coolidge wrote: "It must not be imagined that the appointment of Osgood and Bôcher produced any immediate or revolutionary changes in the mathematical conditions at Cambridge. Fortunately such changes were not needful."[17] Coolidge might have disagreed, therefore, with Birkhoff's use of the word "revolutionary"—and indeed there were no mass firings of the existing faculty or burnings of outmoded textbooks at Harvard—but Birkhoff's term captured the deep nature of the change. The transformation spearheaded by Osgood and Bôcher was profound and led to Harvard's emergence as a mathematical powerhouse within a very short time.

1.2.2 Consolidation and growth: 1900–1933

Parshall and Rowe characterize the period 1900–1933 as one of consolidation and growth of the newly established American mathematical research community. As this period began, the University of Chicago, Harvard and Princeton established themselves as the "Big Three" in mathematics research in the country. According to Parshall and Rowe:

> With these three mathematics departments and several other reasonably sound doctoral programs around the country to choose from, aspiring young American mathematicians no longer felt compelled to go abroad for their training as they had throughout much of the period 1876–1900. Almost without exception, the leading figures of the following period had either studied or taught at one or more of the "Big Three" institutions. These schools

---
[16] See Birkhoff, 1988, p. 15.
[17] See Coolidge, 1930.

were clearly filling the institutional void that had existed prior to 1876.[18]

The University of Chicago established a reputation for research in modern algebra and the calculus of variations. Harvard, under the influence of Osgood and Bôcher, developed an international reputation in analysis and a level of strength in geometry due to J.L. Coolidge. These two schools, combining strong undergraduate and graduate programs in mathematics, played a big part in producing a new generation of well-trained research mathematicians. Led by Henry Burchard Fine, Princeton recruited a distinguished faculty in 1905, including Luther Eisenhart, Oswald Veblen, Gilbert Ames Bliss and John Wesley Young. Then, in 1908, Bliss left for the University of Chicago and Young for the University of Illinois and they were replaced by G.D. Birkhoff, who moved to Harvard in 1912, and J.H.M. Wedderburn. Princeton's reputation was based on the accomplishments of these and other recruited faculty members—it did not develop its own strong graduate program until the 1930s. The efforts of the "Big Three" combined with those of several other schools caused the newly emerged research community to grow, and their research results continued to put the American community on the mathematical map of the world.[19]

1.2.3 Transformation from emergence to consolidation and growth—The bridge years

For examining the research and teaching activities of mathematicians in the United States, the years that formed a bridge between the two periods—1892 to 1907—take on particular significance. In 1892, the University of Chicago opened its doors with E.H. Moore at the head of its mathematics department. At Harvard, Osgood and

---
[18] See Parshall and Rowe, 1994, p. 439.
[19] See Parshall and Rowe, 1994, Chapter 10, entitled "Epilogue: Beyond the Threshold" for further details.

Bôcher had returned from Germany and were establishing themselves in the department. These American mathematicians and a few others started publishing high quality research that flowed into the mainstream of European mathematical interests of the time. This handful of American research mathematicians then produced a second generation of world class research mathematicians. It was when this second generation received their doctorates and began research life, that the transformation in research had been definitively achieved. Until this second generation was on its way, there was no reason to be assured that the mathematicians of the first generation represented more than an isolated phenomenon. G.D. Birkhoff, possibly the most influential mathematician of this second generation, received his Chicago Ph.D. in 1907. In that same year, Osgood published his important treatise *Lehrbuch der Funktionentheorie* (Textbook on the Theory of Functions) which would be the standard for teaching the theory of functions in the United States for years to come.

At Harvard, Osgood played a significant role in the later years of the emergence of the American mathematical research community. He also proved influential in the period of consolidation and growth. Osgood's long career at Harvard spanned this entire period. It is, therefore, of interest to examine his impact during the bridge years from the period of emergence to the period of consolidation and growth. He contributed to consolidation through continuing mathematical research, through leadership in the American Mathematical Society, and through his efforts to build an outstanding mathematics department at Harvard. He and Bôcher contributed to growth through the nurture of their students, both doctoral and other, through writing textbooks that were used for many years, and through their dialogue with the mathematical community on the nature of mathematics education. Both Osgood and Bôcher were exemplary in taking seriously research, teaching and service to their institution and to the American mathematical community as a whole. As a result their influences became deeply woven into the fabric of that community.

1.3 The New Complete Mathematics Professional

William Fogg Osgood lived his professional life as an educator, a mathematics researcher and a participant in a community of mathematicians. Osgood's 18th- and 19th-century predecessors at Harvard had emphasized teaching mathematics to undergraduates; his 20th-century successors, however, embraced the dual objectives of teaching and research that he and Bôcher brought back with them from Germany. Osgood's career at Harvard thus provides a leading example of the transformation that took place in American mathematics. He was an influential and activist member of the generation that brought about the emergence of the American mathematical research community and began to shape the professional structure that still exists over a century later, combining research, teaching and participation in a larger professional community. Moreover, the high quality of his research achievements and their close match with mainstream European mathematical interests of the time played a part in attracting the positive attention of European mathematicians to the new community, giving it a much desired legitimacy.

1.4 Membership in the New Community of Mathematicians

The American Mathematical Society formed under its original name, the New York Mathematical Society, in 1888 while Osgood was studying in Germany. Thomas S. Fiske, Edward L. Stabler and Harold Jacoby, graduate students at Columbia College, proposed the formation of the society "for the purpose of preserving, supplementing, and utilizing the results of their mathematical studies." In addition to mathematical discussion, the trio envisaged a forum for research where "original investigations to which members may be led shall be brought before the society at its meetings."[20] In 1891, the year after Osgood began teaching at Harvard, the Society began to publish its Bulletin and to recruit members from outside the New York area, including William Byerly and Frank Nelson Cole at Harvard. As a result, membership climbed from 23 to 210. In 1894, while Osgood was an assistant professor at Harvard, the group recognized its national character and became the American Mathematical Society. In 1896, membership grew to 275 and the organization of the Chicago section was authorized. Thus, just at the time when Osgood was beginning to produce original research, he had an organized community of mathematicians to welcome it and another publication in which to publish it, the Bulletin of the American Mathematical Society.

The American Journal of Mathematics (1878–present) and the Annals of Mathematics (1884–present) were two other fairly new American publications in existence at the time. The AJM was a high quality forum for mathematical research, about half of which came from the Department of Mathematics at the Johns Hopkins, where it was published, and the other half from Europe and America. Ormond Stone started the Annals of Mathematics at the University of Virginia, catering to a less research-oriented audience.[21]

Osgood both benefited from and contributed to the growing American Mathematical Society. From 1892 until publication of the AMS *Transactions* began in 1900, Osgood published seven items in the *Bulletin*, as compared to three in the *American Journal of Mathematics*, one in the *Annals of Mathematics* and four in German publications. Maxime Bôcher was instrumental in founding the *Transactions* in 1900, and Osgood published his proof of the

---
[20] See Archibald, 1938, p. 4.
[21] See Parshall and Rowe, 1994, pp. 91, 411–412.

Figure 1.3: Frank Nelson Cole

Riemann mapping theorem (see Chapter 4) in the new journal that year.[22] This paper immediately attracted the attention of key European mathematicians and so immediately highlighted the new publication. It was Bôcher who suggested the name "Transactions"—a name that was designed to smooth over difficulties with the AJM which was worried about the new competition in publishing research. But no one could object to any society publishing its "transactions." Bôcher served as its editor in chief in 1908–1909 and 1911–1913, with Osgood taking over for the year 1910.

Osgood also participated in the growing mathematics community by presenting an overview of his field of research, the theory of functions, in the second series of AMS Colloquium Lectures in 1898 and again by giving Colloquium Lectures in 1913 at Madi-

---

[22] Osgood's recollections of the founding of the *Transactions* are reproduced in Archibald, 1938, p. 58.

son with Leonard Eugene Dickson. At an even more official level, he served terms as AMS vice president in 1903 and president in 1905–1906. Osgood also participated in the operation of the AMS through his membership on a committee with L.E. Dickson, H.G. Fine, E.R. Hedrick, P.F. Smith and H.S. White to consider the Society's future financial needs. Their 1920 recommendations were acted on—dues were increased, the *Bulletin* subscription price was raised and a drive for new members resulted in 110 applications. In 1921, along with H.W. Tyler and D.E. Smith, Osgood was appointed to a committee to incorporate the AMS.[23]

Osgood's role in the mathematical community was largely national in nature, as opposed to international. Unlike Bôcher, who gave an address at the 1912 International Congress of Mathematicians and spent a year as exchange professor at the University of Paris, Osgood did not make himself a tangible physical international presence. He could, nevertheless, count himself as a member of an international community of mathematicians, joining in the mathematical research dialogue of the time that included Axel Harnack, Henri Lebesgue, Émile Borel, Henri Poincaré and others, corresponding with both Felix Klein and David Hilbert, contributing an important section on analytic functions to Klein's *Enzyklopädie der Mathematischen Wissenschaften,* and serving as one of three American commissioners of the International Commission on the Teaching of Mathematics.

Osgood's predecessors at Harvard, men such as William Byerly, B.O. Peirce and James Mills Peirce, were not members of national or international communities of mathematicians to any significant extent. Their primary roles were as teachers and administrators at Harvard. With Osgood's career, the transformation becomes evident. Osgood attended and read papers at national professional meetings. He served his professional organization, the AMS, in various offices, including the highest, and worked on its committees.

---

[23]See Archibald, 1938.

Figure 1.4: E.H. Moore

He participated as a member of an American mathematical community in ways that his predecessors did not and his successors still do. He and his Harvard mathematics colleagues after him worked in a mathematical world that could call itself a professional mathematics community.

1.5 A Responsible Educator in a New Style

At Harvard, and in the larger mathematical community, Osgood played a role as a concerned and responsible educator. In general, he demonstrated interest in issues of undergraduate education, while his colleague Bôcher concentrated on graduate education. Osgood also played some part in the larger mathematical education community, both in the United States and abroad. An example already mentioned was his work on the International Commission on the Teaching of Mathematics. (Bôcher chaired the American subcommittee on graduate work.) In addition, in 1902, Osgood, H.W. Tyler, T.S. Fiske, J.W.A. Young and A. Ziwet developed mathematics requirements for college entrance, which were published in the *Bulletin of the AMS* in 1903. Osgood made his vision for the teaching of calculus in the United States known through his 1907 AMS Presidential address in which he advocated both rigor and more emphasis on the physical applications of calculus.

At Harvard, Osgood worked to realize his vision for undergraduate mathematics education. His calculus and other textbooks were used by generations of students and professors. These textbooks demonstrated his deep regard for rigor and the methods of modern analysis, combined with a concern for finding the level of rigor students were capable of handling in a particular course. They were rich in examples and topics from physics. At a more advanced level, the two volumes of his treatise *Lehrbuch der Funktionentheorie* (Textbook on the Theory of Functions) became the standard books on functions of one and several complex variables, and represented one of the most influential accomplishments of his career. In order to train students, high quality textbooks are required, and in this respect Osgood made an important contribution as an educator to the consolidation and growth of the new community.

Osgood was not, however, an inspiring teacher of graduate students. It was Maxime Bôcher's role to supervise many of the doctoral dissertations of mathematics students at Harvard during the earliest years of the century. And Harvard could in no way compete with University of Chicago in numbers of doctorates produced. Still, together the two men, the Harvard team, were instrumental in shaping an educational framework for mathematics that supported the deepening research tradition.

1.6 Top Mathematical Researcher

Osgood left his most impressive mark on the Harvard and American mathematical communities through his mathematical research. Later chapters will examine three major papers, but Osgood also published other significant work in complex analysis, the calculus of variations and differential equations. In a 1938 examination of the tradition of analysis in the United States, G.D. Birkhoff concluded that functional analysis could be traced to E.H. Moore; potential theory, Fourier series, boundary-value problems and ordinary linear differential equations to Maxime Bôcher and functions of one or more complex variables to Osgood[24]

The first world-class research paper Osgood published, "Non-Uniform Convergence and the Integration of Series Term by Term," earned him a place in the international development of analysis, in lines of research that included Axel Harnack, Henri Lebesgue and Émile Borel. Osgood's predecessor in mathematics research at Harvard, Benjamin Peirce, did high-level research, but his most important mathematical work was largely unknown until after his death. He was not to any very great extent a part of a larger mathematical research community, although he was an extremely active member of a general scientific community. This 1897 paper of Osgood reflected the transformation taking place—it announced that the emergent American mathematical community would be a research community and that its research would be at a high level.

Osgood's 1900 proof of the general case of the Riemann mapping theorem reiterated that announcement in a forceful way. His proof corrected and extended work that had been done by Harnack. It used ideas of Henri Poincaré in ways Poincaré had not envisaged.

---

[24]See Birkhoff, 1938, p. 296. R.C. Archibald summarized some of the research of Osgood and Bôcher in Archibald, 1938, and included complete bibliographies. J.L. Walsh wrote a biographical sketch of Osgood which contains some of the high points of Osgood's research. See also Walsh, 1989. G.D. Birkhoff wrote an extensive analysis of Bôcher's work entitled "The Scientific Work of Maxime Bôcher." See Birkhoff, 1919, which also contains a bibliography.

Once again, Osgood proved himself a talented member of the world community of research mathematicians; his reputation—and by extension that of the newly emergent American mathematical community of 1900—was solidifying. This process continued in 1903 with the publication of Osgood's paper exhibiting a Jordan curve of positive area. A logical outgrowth of the consideration of interesting simply connected regions of the plane that began with his work on the Riemann mapping theorem, Osgood's result answered a question posed by Camille Jordan, and again brought him into an international mathematical dialogue.

An examination of Osgood's early research results demonstrates that the U.S. academic community had come to support high-level mathematical research. Moreover, that research attracted the attention of top European mathematicians—attention which was important to a young mathematical community struggling to make its name. Osgood's research, and that of Maxime Bôcher, defined a tradition of research at Harvard that the work of their undergraduate student and later colleague, G.D. Birkhoff, would further exemplify.

1.7 Osgood's Harvard

The year 1907 witnessed the publication of the first volume of Osgood's *Lehrbuch der Funktionentheorie* and of his first calculus textbook, his presidential address before the American Mathematical Society, and the Ph.D. under E.H. Moore at the University of Chicago of a member of Harvard's next generation of research mathematicians, G.D. Birkhoff. Thus 1907 was a landmark year in the transformation of American mathematics as manifested at Harvard. Osgood was a researcher, a teacher and an active member of an American mathematical research community to an extent like no other member of the Harvard mathematics department before him.

Osgood and Bôcher were the leading Harvard members of a small first generation of mathematicians in the United States who defined their positions in this new way. Their conception of a professional mathematician was quite similar to that of mathematicians in the United States over a century later.

G.D. Birkhoff was the first of Harvard's next generation of top research mathematicians, joining the faculty in 1912.[25] Nationally, the country's mathematical leadership began to pass from E.H. Moore, Osgood, Bôcher and H.B. Fine at Princeton, to men like Birkhoff at Harvard, Gilbert Ames Bliss at University of Chicago and Oswald Veblen at Princeton, as Princeton joined the University of Chicago and Harvard in playing a major leadership role in the community.[26] Birkhoff was later joined by Oliver D. Kellogg in 1919, by Harvard Ph.D. Joseph L. Walsh in 1921, by his own former doctoral students H.C. Marston Morse in 1926 and Marshall H. Stone in 1927. These men carried Harvard through the remainder of the American mathematical research community's period of consolidation and growth (1900–1933).

Harvard's most prominent faculty member in mathematics and science entering the period of emergence of the American mathematical community was Benjamin Peirce. The retirement of William Byerly and the death of B.O. Peirce in 1913 marked "the end of Benjamin Peirce's influence on mathematics at Harvard."[27]

Benjamin Peirce's scientific circle at Harvard entered the period of emergence of the American mathematical research community and became transformed into G.D. Birkhoff's Harvard mathematics department enjoying a rich period of growth, an international reputation and an ever stronger research tradition in analysis. William Fogg Osgood was an agent of this transformation in two senses of

---

[25] J.L. Coolidge and E.V. Huntington joined Osgood and Bôcher on the faculty at Harvard before Birkhoff—Coolidge in 1902 and Huntington in 1901. Both studied in Germany and eventually served terms as President of the Mathematical Association of America.

[26] See Birkhoff, 1977, p. 55.

[27] See Birkhoff, 1989, p. 27.

the word. On one hand, coming at the end of the period of emergence, he was a passive agent of the transformation, his work a product of the foundations laid by Sylvester, Klein, Moore, Eliot, Peirce and others. On the other hand, Osgood came into his professional maturity at the beginning of the following period, and became (and then remained throughout the entire 1900–1933 period) an active agent of the transformation, one of the key players, particularly at Harvard, in shaping the newly emerged community of mathematicians into the form it still generally maintains today. He acted as a *responsible educator*, though certainly not a perfect one, who both benefited from and set new traditions and standards at Harvard and nationally. His leading role, that of *top mathematical researcher*, was both a factor in legitimizing the mathematical community as a *research* community, and in developing a strong research tradition in analysis. An examination of Osgood's career is therefore, both a Harvard epilogue to the period of emergence of a mathematical community, and a prologue to a larger historical examination of the American mathematical community in the early 20th century.

1.8 Osgood's Research and Teaching—A Taste of Chapters to Come

William Fogg Osgood's research, teaching and participation in a mathematical community provide a Harvard case study of the transformation of the American mathematical community in the years 1892 to 1907.

In Chapter 2 the stage is set for Osgood, and indeed for his colleague Bôcher, by considering some of the components of the mathematical foundation for the transformation laid at Harvard by their predecessors, men such as John Farrar, Nathaniel Bowditch, Benjamin Peirce and Frank Nelson Cole. It also describes the in-

1. Transformation in the Air

stitutional foundation set in place by Harvard President Charles William Eliot. These men and others formed the mathematical environment in which Osgood and Bôcher were trained as undergraduates and later welcomed as faculty members.

Chapters 3 through 5 analyze three of Osgood's most important research papers. Along with other significant work, these three papers made Osgood, for a while, a premier mathematical researcher in the United States, with his work forming part of the European mathematical mainstream. Each of Chapters 3 through 5 contains a description of the historical and mathematical context into which Osgood's results fit, a technical analysis of those results, and some of the subsequent reaction to the results, including indications of how they affected future research.

Osgood's first major research paper, *Non-uniform Convergence and the Integration of Series Term by Term*, published in 1897, is the topic of Chapter 3.[28] The result can be viewed as a precursor to Lebesgue's 1908 dominated convergence theorem. Chapter 4 delves into his paper *On the Existence of the Green's Function for the Most General Simply Connected Plane Region*, which completed the first proof of the Riemann mapping theorem for the most general simply connected plane region.[29] These are undoubtedly two of Osgood's most important papers. A third important paper is striking for the non-intuitive nature of its subject matter. The paper, *A Jordan Curve of Positive Area*, appeared in 1903 and is analyzed in Chapter 5.[30]

Chapter 6 examines Osgood's educational activities, in particular his approach to teaching undergraduates, as demonstrated by his ideas about calculus instruction reform and some of his long-lived textbooks. Maxime Bôcher's teaching at Harvard complemented Osgood's in many ways. Osgood wrote Harvard's calculus

---

[28] See Osgood, 1897a.
[29] See Osgood, 1900.
[30] See Osgood 1903a.

books, while Bôcher wrote the higher-level algebra book, for example. Moreover, unlike Osgood, Bôcher directed a significant number of Ph.D. dissertations, helping ensure that Harvard contributed its share of Ph.D.s to the growing American mathematical community. Chapter 6, therefore, includes a section about Bôcher's textbooks, teaching and doctoral students.

This book joins together the pieces of the picture of William Fogg Osgood as mathematical researcher, educator and member of a larger professional community at Harvard and in the United States, exemplifying the transformation of American mathematics in the years around the turn of the century. At Harvard, this transformation of mathematical life had long roots. Osgood and Bôcher followed in the footsteps of Isaac Greenwood who became the first Hollis Professor of Mathematics and Natural Philosophy in 1727, of John Farrar who translated then-modern French texts for use at Harvard in the early 1800s, of self-taught scientist Nathaniel Bowditch, and of Benjamin Peirce who served as a role model for mathematical research. Osgood and Bôcher benefited from the educational vision and ideals of Harvard President Charles Eliot. They built on the accomplishments of their early teachers at Harvard, James Mills Peirce, William Byerly and Benjamin Osgood Peirce. By the time Osgood and Bôcher finished their years of study at Harvard, the mathematical foundation for the transformation of American mathematics, as manifested at Harvard, had been put into place.

Chapter 2

Before Osgood and Bôcher—Signs of a Transformation to Come

The transformation of American mathematics in the latter part of the nineteenth century happened relatively quickly—a mathematical foundation had already been built. This foundation took form in the century from 1776 to 1876, the first period in the history of American mathematics. Parshall and Rowe observe that, during this first period:

> ...the field evolved not as a separate discipline but rather within the context of the general structure-building of American—as opposed to colonial—science. The colleges formed a primary locus of scientific activity, but, by and large, they did little to encourage the pursuit of research for the advancement of science. At the same time, the concept of research in American science—as in other academic disciplines—emerged as scientists looked toward

Europe as their model and measured themselves against the yardstick of European scientific achievement.[1]

Historian of education Laurence Veysey wrote about the emergence of the concept of research in American higher education in general:

> In the 1870's research played no important role in American higher education. Indeed, at that time the idea of a formal academic career was still in its infancy... Around 1880 a definite change occurred. It then began to be believed—whether rightly or not—that most of the "bright young men" were going into science... it was in 1880 that Willard Gibbs first was paid a salary by Yale. Ten years later research had become one of the dominant concerns of American higher education.[2]

In particular, research became one of the dominant concerns of the faculty involved in mathematics at institutions of higher education. With increasing numbers of Americans inspired in great measure by European models, the period 1876–1900 saw the emergence of an American mathematical research community. By 1907 the country had a thriving community of mathematicians working on some of the most important research topics of the day.

The mathematics community at Harvard shared this awakened concern for research. Moreover, the members of that community gave due attention to the education and training of future generations of mathematicians who would then sustain the new research tradition. This chapter examines some of the foundation laid at Harvard for the resulting transformation of American mathematics, especially those sections of the foundation that influenced the mathematical education of William Fogg Osgood.

---
[1] See Parshall and Rowe, 1994, p. xiii.
[2] See Veysey, 1965, pp. 174–175.

2.1 Building a Mathematical Foundation: 1727–1869

2.1.1 The Hollis Professors at Harvard—A traditional British curriculum and non-specialization

The Hollis Professorship of Mathematics and Natural Philosophy was the first professorship of a "profane" topic at Harvard. Its first occupant, Isaac Greenwood (1802–1745), held the post from 1727 until his dismissal in 1738. J.L. Coolidge, a much later successor of Greenwood as a professor of mathematics at Harvard, alluded to the reason for the dismissal in an essay on the history of the department. While Greenwood "had the merit of orthodoxy, he had the failing of being a confirmed drunkard."[3] Greenwood, a Harvard graduate who had also studied in England, taught algebra to his students. His course appeared to be based on writings of 17th-century scholar John Wallis who had been Savilian professor of geometry at Oxford for over fifty years. Historians David Eugene Smith and Jekuthiel Ginsburg mention two works in particular—*Arithmetica Infinitorum* published at Oxford in 1655 and *Opera Mathematica*, published in 1693–1695. Greenwood taught solving quadratic and "cubick" equations and "Dr. Halley's Theorems for Solving Equations of all sorts" and at least mentioned the works of Euclid, Apollonius and Archimedes.[4] He published an arithmetic text anonymously in Boston in 1729 the authorship of which was later established by means of a contemporary advertisement in a newspaper.[5] According to Greenwood's entry in the *Dictionary of Scientific Biography*, he "seems to have taught Newtonian fluxions" (vol. 5, 1972, pp. 519–520). Greenwood, more than was usual for the time, specialized in teaching mathematics. As was also common in Europe, American professors tended to teach mathematics and a variety of sciences, and sometimes other subjects as well. Greenwood taught a traditional British-style mathematics curricu-

---
[3] See Coolidge, 1930, p. 248.
[4] See Smith and Ginsburg, 1934, pp. 20–21.
[5] See Smith and Ginsburg, 1934, pp. 38–39.

lum, without the influences of continental mathematics that would increasingly appear in the teachings of his later successors.

An accomplished astronomer who was one of the founders of the American Academy of Arts and Sciences, John Winthrop (1714–1779), took up the Hollis chair in 1738 at age 24. He was a Harvard College graduate and a former student of Greenwood. A 1764 letter written by Winthrop explained his wide-ranging duties as professor of mathematics and natural philosophy:

> My province in the College is to instruct the students in a system of Natural Philosophy and a course in Experimental, in which is to be comprehended Pneumatics, Hydrostatics, Mechanics, Statics, Optics, etc.; in the elements of Geometry, together with the doctrine of Proportion; the principles of Algebra, Conic Sections, Plane and Spherical Trigonometry, with general principles of Mensurations of Planes and Solids; in the principles of Astronomy and Geography, viz. the doctrine of the sphere, the use of the globes; the calculations of the motions and phenomena of the heavenly bodies according to the different hypotheses of Ptolemy, Tycho Brahe, and Copernicus, with the general principles of Dialling; the division of the world into various kingdoms, with the use of the maps, and sea charts; and the arts of Navigation and Surveying.[6]

Winthrop established the country's first experimental physics laboratory and re-built Harvard's collection of scientific instruments following a 1764 fire that destroyed Harvard Hall[7]. He published numerous astronomical observations in the *Philosophical Transactions of the Royal Society*. His varied professorial duties showed a characteristic level of non-specialization for his time. Like Green-

---

[6]See Block, 1933, p. 660. See also Smith and Ginsburg, 1934, pp. 54–55.
[7]See *Dictionary of Scientific Biography*, v. 14, 1976, p. 452.

wood's curriculum, Winthrop's mathematical teachings had a primarily British lineage.

The mathematics curriculum at the colonial colleges mirrored the centuries-old British curriculum based largely on Euclid, i.e., an axiomatic approach with a heavy emphasis on geometry, and the addition of some Newtonian mechanics. But changes were happening on the Continent that some Americans, including John Farrar at Harvard and his contemporary Nathaniel Bowditch, soon started to notice.

2.1.2 John Farrar, the Fifth Hollis Professor—A French approach to mathematics

John Farrar (1779–1853), who accepted the Hollis Professorship after Nathaniel Bowditch declined it, worked to bring the French approach to mathematics to the United States. This approach favored algebra over geometry. Its style of teaching was more exploratory and problem-solving in nature, in contrast to the British-style axiomatic development of mathematics. In the new style, students would learn mathematics by examining situations in which the mathematics applied.

Following the War of Independence, the United States had its political independence from Great Britain, but still felt the influence of British mathematics. Americans imported books from Britain and brought out American editions of British texts in mathematics. In 1801, for example, Samuel Webber prepared a two-volume text entitled *Mathematics, Compiled from the Best Authors and Intended to Be a Text-book of the Course of Private Lectures on These Sciences in the University at Cambridge*.[8] Webber was the fourth Hollis Professor of Mathematics and Natural Philosophy at Harvard. According to Florian Cajori, the volumes "embraced the

---
[8] See Webber, 1801.

subjects of arithmetic, mensuration of solids, gauging, heights and distances, surveying, navigation, conic sections, dialing, spherical geometry, and spherical trigonometry."[9] The publication of Webber's text nearly coincided with the 1802 strengthening of admission requirements for Harvard. The revised requirements included some knowledge of arithmetic.[10]

American mathematics took its lead from British mathematics, but British mathematics had stagnated following Newton's accomplishments. In the opening years of the 19th century, British mathematicians such as Charles Babbage, George Peacock and John Herschel recognized this and took steps to correct the situation, one of which was the formation of the Cambridge Analytical Society in 1812. The Society met regularly to discuss mathematics in the new French analytic style. By the 1820s, a more "general spirit of reform had overcome Cambridge," and continental mathematics eventually infiltrated the curriculum at Cambridge.[11]

Around the same time, fifth Hollis Professor John Farrar started translating continental texts to use in his courses, rather than British ones like his predecessor Webber. Farrar graduated from Harvard in 1803, received his master's degree in 1805 and was appointed Hollis Professor two years later in 1807. He held the chair until 1836. Farrar's Americanized version of Silvestre Lacroix's, *Traité élémentaire d'Arithmétique* (1797, Elementary Treatise of Arithmetic) appeared in 1818. His translation of Adrien-Marie Legendre's *Éléments de Géométrie—Avec des Notes* (1794, Elements of Geometry—with notes) came out in 1819, followed in 1820 by an edition of Lacroix's *Traité élémentaire de Trigonométrie rectiligne et sphérique* (1798, Elementary treatise of plane and spherical trigonometry). Farrar, together with George Emerson, based their *First Principles of Differential and Integral Calculus* on the

---

[9] See Cajori, 1890, p. 60.
[10] See Parshall and Rowe, 1994, p. 3.
[11] See Parshall and Rowe, 1994, p. 7. For a discussion of the roles of British and French influences on mathematics in the U.S., see Parshall and Rowe, 1994, pp. 2–15.

2. Before Osgood and Bôcher 35

work of Étienne Bézout. It was published by 1824. *First Principles of Differential and Integral Calculus* introduced the notation of Leibniz—as opposed to that of Newton—to the United States. But this did not seem to be the most important thing to Farrar. Rather he praised Bézout's plainness and brevity as appropriate to American students with little time to devote to such studies.[12] His translations found an audience beyond Harvard—they were used at the United States Military Academy at West Point and other colleges.

Farrar's efforts to bring Continental-style mathematics to Harvard formed an early part of the foundation for eventual research-level mathematics at Harvard. His works were still at an elementary level, and were translations of material that was already fairly old, but they were more appropriate than previous texts to the serious preparation of beginning students of mathematics. In 1830, during Farrar's tenure, freshmen at Harvard read Legendre's *Geometry*, and studied algebra and solid geometry. Sophomores read trigonometry, topography and calculus. All juniors studied natural philosophy, mechanics, electricity and magnetism. Seniors took natural philosophy and optics.[13] Graduate education was still years in the future.

2.1.3 Nathaniel Bowditch—Making available the "new" French methods

One man whose influence was important to mathematics and astronomy at Harvard never joined the faculty there. Nathaniel Bowditch (1773–1838) was born in Salem, Massachusetts and spent many years at sea, eventually commanding a ship on a voyage to Sumatra.[14] Bowditch was a self-taught man who learned Latin in

---
[12] See Parshall and Rowe, 1994, p. 8.
[13] See Parshall and Rowe, 1994, p. 17.
[14] See Reingold, 1964, p. 11.

1790 in order to read Newton's *Principia*, and German at the age of 45 in order to read current scientific literature. His studies were aided by his access to a collection of books which had belonged to Robert Kirwan, an Irish chemist. This collection included the *Philosophical Transactions of the Royal Society*, as well as works in mathematics, physics, astronomy and chemistry.[15] Long trips at sea afforded him ample time for study. He became familiar with the Leibnizian approach to calculus through study of Lacroix's *Traité élémentaire de calcul différentiel et de calcul intégral*. He put his seafaring knowledge to use writing *The New Practical Navigator*, published in 1802, which became and, in later editions, remains a standard reference work for seamen. The 2002 edition is still in print.[16] He put his knowledge of French mathematics to use in his translation with commentary of Pierre-Simon Laplace's *Traité de mécanique céleste* (Treatise on Celestial Mechanics).

Bowditch was highly respected by those with an inclination to science, and received invitations to teach at Harvard, the U.S. Military Academy at West Point and the University of Virginia but turned down the offers. In 1803 Bowditch was given an honorary M.A. from Harvard but in 1806 he declined the Hollis Professorship that was subsequently accepted by John Farrar. Even given the academic standards of the time, it says much about Bowditch's reputation and abilities that Harvard offered this post to a self-educated man from a poor family with no academic experience. Bowditch's refusal to teach at Harvard did not preclude a long and close association of the Bowditch family with the college. His son, Henry Ingersoll Bowditch was Benjamin Peirce's classmate at Har-

---

[15] See Parshall and Rowe, 1994, p. 9.

[16] For biographical sketches of Bowditch, see Parshall and Rowe, 1994, pp. 8–12, Reingold, 1964, pp. 11–14, and Reingold, 1970. Bowditch's son, Henry Ingersoll Bowditch wrote a biographical study of his father to which Parshall and Rowe refer. See his "Memoir" in Pierre Simon de la Place, *Celestial Mechanics*, trans. Nathaniel Bowditch, 4 vols. (New York, Chelsea Publishing Co., 1966) 1:9–168. Of interest to children is the Newberry award winning biographical novel *Carry On, Mr. Bowditch* by Jean Lee Latham (New York, Houghton Mifflin Company, 1955).

2. Before Osgood and Bôcher

vard. Bowditch became a member of the Harvard Corporation, the college's governing body, in 1826.[17]

The offer to teach at the University of Virginia was made by Thomas Jefferson in a letter from Monticello on 26 October 1818 which commented on Jefferson's own mathematical ability, opening with the sentences:

> I have for some time owed you a letter of thanks for your learned pamphlet on Dr. Stewart's formula for obtaining the sun's distance from the motion of the moon's Apsides; a work however, much above my Mathematical stature. This delay has proceeded from a desire to address you on an interest much nearer home, and on the subject of which I must make a long story.[18]

Jefferson described the college being founded in Charlottesville, the climate there, the salary and facilities, the near guarantee of tenure for life and the desire to hire only men of the first order for the faculty. Then came the actual invitation: "We are satisfied that we can get from no country a professor of higher qualifications than yourself for our Mathematical department, and we entertain the hope and great anxiety that you will accept of it." On 4 November, Bowditch sent his refusal to Jefferson from Salem pleading important financial trusts "undertaken for the children of an ancient merchant late of this Town, and which upon principles of honor cannot be resigned but with my life, bind me to the powerful hands of *interest* to the place of my nativity..." He also offered as excuse the delicate health of his wife and, what may have been the deciding factor: "Considerable time is left to me [after discharge of various obligations] to devote to those studies which have been the delight of my leisure

---
[17]See Birkhoff, 1989, pp. 4–5.
[18]Letter from the Nathaniel Bowditch Papers of the Boston Public Library and contained in Reingold, 1964, pp. 20–23.

hours."[19] In this correspondence, Bowditch showed no evidence of dissatisfaction with his position as an academic outsider. The position of college professor had not yet become the obvious one for the learned person who wanted to pursue scholarly interests. Bowditch needed his "leisure hours" for his studies.

Figure 2.1: Nathaniel Bowditch by Chester Harding

Turning down these prestigious job offers also had financial benefits for Bowditch. In 1804, after retiring from the seafaring life, he became president of a marine insurance company in Salem; then in 1823 he took a position as an actuary at the Massachusetts Hospital Life Insurance Company.[20] These jobs put Bowditch in a financial position that allowed him to pay for the publication of one of his

---

[19]Letter from the Jefferson Papers of the Library of Congress and contained in Reingold, 1964, p. 23.
[20]See Parshall and Rowe, 1994, p. 9.

most important works, a translation with commentary of Laplace's *Traité de mécanique céleste*. In this work, Bowditch supplied many details omitted in the original text, incorporated later results and gave credit to others that Laplace did not give. It was perhaps because of this assignment of credit that there is no evidence of any response from Laplace to communications from Bowditch about the work.[21] Four volumes of the "translation" of the *Mécanique céleste* were eventually published in 1829, 1832, 1834 and 1839, the last posthumously.

Bowditch was already a respected figure even before publication of his version of the *Mécanique céleste*. In 1811 he published a paper on the 1807 meteor explosion over Weston, Connecticut in *Nicholson's Journal*, and in 1815 he published a paper in the *Memoirs of the American Academy of Arts and Sciences* on the motion of a pendulum suspended from two points. He had also already been made a member of the Royal Society.[22] But the *Mécanique céleste* was singularly important in that it made more portions of the French mathematics available in English to American students. He made a further contribution to American mathematics after his death when his heirs opened to public use his private mathematical library of 3,000 volumes.[23]

Nathaniel Bowditch thus contributed to the foundation for the transformation of mathematics at Harvard through his stature as a man of science, through his work on the *Mécanique céleste* and through his library. Bowditch made these contributions to the foundation for the transformation as an exception, somewhat of an outsider. Never serving on the Harvard faculty, not an academic, he nonetheless became aware of and assimilated the changes in approach to mathematics happening on the Continent. In spite of being an outsider, he passed his torch to the following generation at Harvard in the person of Benjamin Peirce (pronounced like "purse").

---

[21] See Reingold, 1970, p. 368.
[22] See Reingold, 1970, p. 368.
[23] See Bruce, 1987, p. 39.

The mathematical formation of Benjamin Peirce was another major contribution Bowditch made to the transformation. Peirce would become another type of exception—a mid-19th-century American who did original research in mathematics.

2.1.4 Benjamin Peirce—An American mathematical researcher

Benjamin Peirce (1809–1880) directly benefited from the mathematical tutelage of both Farrar and Bowditch.[24] Farrar was a mentor during Peirce's studies at Harvard, and Bowditch was an influence beginning from the younger man's childhood years when Peirce attended school with Bowditch's son, Henry Ingersoll Bowditch. Peirce dedicated his textbook *Analytic Mechanics*: "To the cherished and revered memory of my master in Science, Nathaniel Bowditch, the father of American geometry..."[25]

Peirce was born in Salem, Massachusetts in April 1809, the namesake of his father Benjamin Peirce who himself had graduated from Harvard in 1801. Benjamin the elder held the post of librarian at Harvard during the last years of his life, served as a member of the Massachusetts state legislature, and wrote a history of Harvard that was published posthumously.

Benjamin the son began his studies at Harvard in 1825, while Farrar held the Hollis Professorship. By this time, Farrar's translations and adaptations of the works of continental mathematicians had already appeared; presumably Peirce was among those to benefit from their availability. He was further aided in his mathematical training through his assistance to Bowditch with the translation of the *Mécanique céleste*—he corrected proof sheets over a ten-year period. Thus Peirce received a more French-influenced, more modern

---

[24]Biographical sketches of Benjamin Peirce by Raymond Clare Archibald and Sven Peterson, as well as various reminiscences and the text of Peirce's "Linear Associative Algebra" appear in Cohen, 1980.

[25]See Peirce, 1855.

2. Before Osgood and Bôcher

Figure 2.2: Benjamin Peirce

mathematical education than many of his immediate predecessors and contemporaries. After graduating from Harvard in 1829 he taught at the Round Hill School before being appointed tutor at Harvard in 1831. He received his master's degree in 1833 and was appointed professor of mathematics and natural philosophy. (Farrar still held the Hollis chair so it was unavailable.) An indication of increased specialization, his title was changed to professor of mathematics and astronomy in 1842.

Peirce broke with the translation tradition of his mentors by writing original texts and publishing a substantial amount of contemporary mathematical research, in addition to his astronomical research. R.C. Archibald began his biographical sketch of Peirce with the line: "Mathematical research in American Universities began with Benjamin Peirce."[26] Both as textbook writer and publisher of important original mathematical research, Peirce broke new ground for mathematics in America in the mid-1800s. A list of Peirce's writings, albeit with commentary, occupies an impressive eleven pages of Archibald's biographical sketch. In the 1840s, Peirce wrote "elementary treatises" (textbooks) on plane and spherical trigonometry, sound, plane and solid geometry, algebra, and differential and integral calculus. The preface to his 1855 *A System of Analytic Mechanics* revealed Peirce's determination both to bring the most up-to-date work to the United States and to encourage research. Stating his case with characteristic grandeur, Peirce wrote:

> I have...reexamined the memoirs of the great geometers, and have striven to consolidate their latest researches and their most exalted forms of thought into a consistent and uniform treatise. If I have, hereby, succeeded in opening to the students of my country a readier access to these choice jewels of intellect, if their brilliancy is not impaired in this attempt to reset them, if in their new constellation they illustrate each other and concentrate a stronger light upon the names of their discoverers, and still more, if any gem which I may have presumed to add is not wholly lustreless in the collection, I shall feel that my work has not been in vain. The treatise is not, however, designed to be a mere compilation. The attempt has been made to carry back the fundamental principles of the science to a more profound and central origin; and thence to shorten the path to the most fruitful forms of research.[27]

The *Analytic Mechanics* was an impressive work, but was not suitable for introducing the subject to students. Astronomer Simon Newcomb commented that: "The exposition of dynamical concepts

---
[26] See Archibald, 1925, p. 8.
[27] See Peirce, 1855, preface, in Archibald, 1925, p. 25.

2. Before Osgood and Bôcher 43

in the first forty pages is pleasant reading for one already acquainted with the subject, but that a student beginning the subject could understand it without clearer distinction of definitions, axioms, and theorems seems hardly possible."[28]

Peirce is best known among mathematicians for his 1870 publication *Linear Associative Algebra*. This work was read before the National Academy of Sciences in Washington D.C., privately published by Peirce, then republished posthumously in the *American Journal of Mathematics* with notes and addenda by his son Charles Saunders Peirce in 1881.[29] Benjamin Peirce dedicated the original version to his friends and wrote: "This work has been the pleasantest mathematical effort of my life. In no other have I seemed to myself to have received so full a reward for my mental labor in the novelty and breadth of the results."[30] His work on linear associative algebras, higher systems of complex numbers, grew out of an interest in the quaternions, which William Rowan Hamilton had discovered in 1843, and on which Peirce had lectured at Harvard in 1849. The quaternions demonstrated that an algebra could have all the properties of the real or complex numbers, except that multiplication was not commutative. John Graves and Arthur Cayley then independently discovered the "octonions," an algebraic system in which multiplication is neither commutative nor associative. The number of such "hypernumber" systems rose rapidly—some synthesis and summary of ideas was badly needed.[31] Peirce performed this service for "linear" algebras with complex coefficients and an associative multiplication.

---

[28] See Newcomb, p. 742, in Archibald, 1925, p. 14.
[29] See Peirce, 1881.
[30] Peirce's dedication is from the preface to the original, privately published, 1870 edition and is reproduced in Cohen, 1980.
[31] For historical accounts, see LaDuke, 1983, pp. 147–159 and Kline, 1972, chapter 32.

In *Linear Associative Algebra*, Peirce analyzed the structure of such algebras.[32] Jeanne LaDuke gave a succinct description of Peirce's work:

> He introduced the concepts "nilpotent" and "idempotent" and showed that in any linear associative algebra either there is an idempotent or every expression in the system is nilpotent.[33] Then he demonstrated that in any algebra containing an idempotent, the units (basis elements) may be chosen so that they fall into four disjoint sets defined with respect to the idempotent. This yields what is now called the "Peirce decomposition" of an algebra. Peirce's analysis is followed by a 97-page "investigation of special algebras" in which his goal was to enumerate algebras as described by displaying the multiplication tables of the basis elements.[34]

Peirce's role as a researcher began the establishment of a tradition of mathematical research at Harvard. Before Peirce and continuing to a significant degree through his tenure, Harvard and other colleges did not heartily encourage research interests on the part of the faculty. Scholarly skills were not prerequisites for receiving university appointments. In 1870, Charles Joyce White received a presidential appointment in mathematics at Harvard. Julian Coolidge noted that White's "mathematical knowledge never went beyond the point which a man specially interested in classics needed to reach in order to get a Harvard A.B.; and any natural aptitude he may have had for teaching was successfully extinguished by his service as instructor at the Naval Academy."[35] In fact, there were not enough capable mathematicians in the country for the possession of schol-

---

[32]*Linear Associative Algebra* can be downloaded as a Google eBook free of charge from Google Books.

[33]An element A of the algebra is an idempotent if $A^2 = A$ and nilpotent if there exists a positive integer n such that $A^n = 0$.

[34]See LaDuke, 1983, pp. 148–149.

[35]See Coolidge, 1930, p. 251.

arly skills to be a prerequisite for receiving university appointments in mathematics. Peirce provided his comment on this situation; according to one commentator, Peirce "was once asked what American mathematicians thought of a recent appointment to a professorship in mathematics. Peirce replied that no one had a right to express an expert opinion except himself and one former pupil, Lucien A. Wait, later professor at Cornell."[36] Thus Peirce served as one of the first models of a high-level research scientist, thanks especially to the immediate impact that his work on the orbit of Neptune made in international circles, limited somewhat by the fact that his most important mathematical work, *Linear Associative Algebra*, was not widely known until after his death.

2.1.5 Benjamin Peirce as an educator at Harvard

By 1846, Peirce was firmly established in his career as a scientist and as a fixture at Harvard. The entire faculty at that time numbered twenty, four of them teaching sciences: physics teacher Joseph Lovering, John W. Webster, a chemist who was hanged for murder in 1850, Asa Gray, the nation's leading botanist, and Peirce himself.[37] They were later joined by Swiss naturalist Louis Agassiz, among others.

At Harvard, Peirce did not enjoy the greatest success in his role as teacher. His talents did not lie in the area of education of the average undergraduate, and he had no core of dedicated graduate students to inspire. Sven R. Peterson collected a number of reflections on Peirce as teacher of undergraduates[38] and also gave his own opinion:[39]

---

[36] See Peterson, 1955, p. 92.
[37] See Bruce, 1987, p. 39.
[38] See Peterson, 1955, pp. 93–94.
[39] See Peterson, 1955, p. 93.

Peirce was successful in having mathematics made an elective, first from the Senior year only, but eventually for the whole four years. One compelling motive for this action may have been his intense dislike of teaching any but the most gifted students. When mathematics was made an elective, the students stayed away in droves, and the mathematics department became known as small, difficult, and unpopular. Generations of unhappy students have recorded what they suffered at Peirce's hands, a combination of respect for his enthusiasm and genius with a total befuddlement as to what he was trying to say.

Two obituary notices for Peirce are characteristic of Peterson's evidence of the former's unsuitability for teaching the average undergraduate. Renowned astronomer Simon Newcomb wrote in his obituary:[40]

> As a teacher, he [Peirce] was very generally considered a failure. The general view he took was that it was useless for anyone to study mathematics without a special aptitude for them; he therefore gave inapt pupils no encouragement, and made no attempt to bring his instruction within their comprehension.

An editorial writer for the *Springfield Republican* echoed Newcomb's view:

> Few men could suggest more while saying so little, or stimulate so much while communicating next to nothing that was tangible and comprehensible.[41]

---

[40] See Newcomb, p. 742.

[41] See Peterson, 1955, p. 94. Peterson found this characteristic reaction to Peirce in *Benjamin Peirce, A Memorial Collection*, ed. Moses King, Cambridge, 1881. The obituary ran in the *Springfield Republican* on October 23, 1880.

2. Before Osgood and Bôcher

Two of Peirce's former students attested to his inability to teach most students but also revealed some of his strengths. From his student William E. Byerly (who received Harvard's first Ph.D. in 1873 and later taught Osgood and Bôcher):[42]

> When I knew him later in the class-room, I will not say as a teacher, for he inspired rather than taught, and one's lecture notes on his courses were apt to be chaotic, I always had the feeling that his attitude toward his loved science was that of a devoted worshipper, rather than of a clear expounder. Although we could rarely follow him, we certainly sat up and took notice.

William F. Allen, a member of Peirce's 1851 class was less charitable:[43]

> I am no mathematician, but that I am so little of one is due to the wretched instruction at Harvard. Professor Peirce was admirable for students with mathematical minds, but had no capacity with others.

Peirce's classes were also small. In 1854, a dozen juniors continued their studies of mathematics with him, only four of whom persevered into their senior year.[44] Coolidge observed that Peirce's "great natural mathematical talent and originality of thought, combined with a total inability to put anything clearly, produced upon his contemporaries a feeling of awe that amounted almost to dread."[45]

On the institutional front, in 1847, Harvard established the Lawrence Scientific School to provide a higher level of education in

---
[42] See Byerly, in Archibald, 1925, p. 5.
[43] See Allen, in Cajori, 1890, p.140.
[44] See Parshall and Rowe, 1994, p. 19.
[45] See Coolidge, 1930, p. 248.

science, specifically including mathematics. Inspired by his studies in Germany, Harvard President Edward Everett had asked Peirce to draw up a plan for a graduate program and school of science. But in Peirce's hands the idea became a scientific school with the then-vacant Rumford chair as its centerpiece. After approval of the plan, Everett gave the chair to Eben Horsford on the recommendation of his former teacher, German chemist Justus von Liebig. Horsford managed to obtain a $30,000 donation from textile magnate Abbott Lawrence and Harvard named the new school for him.[46] The new school provided Peirce with an opportunity to advance the study of mathematics at Harvard further. His curriculum included Augustin Louis Cauchy's *Cours d'Analyse* (1821, Course in Analysis), William Rowan Hamilton's quaternions and the mathematical work behind the conjecture of the existence of Neptune (1846) done by John Couch Adams and Urbain Leverrier.[47] Unfortunately, the Lawrence School curriculum enlightened few students as only the most gifted found themselves able to benefit from Peirce's teaching. In 1849, for example, he taught only two students.

There is, therefore, a clear consensus that Benjamin Peirce was a wretched teacher for most Harvard students. But he could inspire the talented student. He would likely have been much more adept at directing and inspiring a dedicated group of graduate students, but the Harvard of the mid-19th century could not yet provide Peirce with such an opportunity. (J.J. Sylvester at Hopkins in 1876 would be the first mathematician to have this opportunity in an American institution.) Peirce's immediate successors at Harvard did not become a first generation of American research mathematicians. As educators, however, Peirce's successors ensured that his influence as a teacher did reach a following generation—Osgood and Bôcher, his successors' students, became his true Harvard mathematical heirs. In addition to Peirce's legacy as a researcher and educator, his ef-

---

[46]See Bruce, 1987, pp. 22, 163.
[47]See Parshall and Rowe, 1994, p. 18.

2. Before Osgood and Bôcher

forts as a scientific activist would also leave their mark on American science and mathematics.

2.1.6 Benjamin Peirce as influential scientist

Americans in the sciences were aware that they did not possess the professional traditions and structures that would facilitate their entry into the top scientific ranks. Some had been to Britain, France and Germany and knew that much work lay ahead of them to make science in America reach European standards of accomplishment. A spirit of scientific activism began to stir. In 1838, soon after their scientific studies in Europe, Joseph Henry wrote to his friend Alexander Dallas Bache:

> I am now more than ever of your opinion that the real working men in the way of science in this country should make common cause and endeavour by every proper means unitedly to raise their own scientific character. To make science more respected at home to increase the facilities of scientific investigators and the inducements to scientific labours.[48]

Benjamin Peirce united with Bache, Henry and others in efforts to accomplish these goals. He became one of the core members of an informal scientific society known as the Florentine Academy or Lazzaroni, the latter a name that originally signified certain Neapolitan ne'er-do-wells. The American Lazzaroni gathered informally in small groups and carried on a correspondence full of bright patter.

The Lazzaroni wanted to advance and protect American science. In 1846, calculations made by French mathematical astronomer U.J.J. Leverrier led astronomers at the Berlin Observatory to the

---
[48] See Reingold, 1964, p. 85 and Bruce, 1987, p. 26.

discovery of the planet Neptune. Pierce, along with American astronomer Sears Cook Walker, jumped into the European scientific discussion, claiming that Neptune was not the planet predicted by Leverrier and that its discovery was an accident. The ensuing debate was highly charged, with the Lazarroni defending Peirce in the interests of the advancement of American science.[49]

The Lazzaroni were led by Bache, affectionately called "Chief" by his followers, who then of course called his wife "Chiefess".[50] Detractors of the group called them "Bache and Company." Bache became superintendent of the U.S. government's Coast Survey in 1843. His friend Joseph Henry, the more gifted scientist of the two and respected for his work in electricity, was the second shaper of the Lazzaroni. He received an appointment as secretary of the Smithsonian Institution in 1846. They were joined by Peirce, whom Bache met in 1842. In 1852, Peirce joined Bache at the Coast Survey. Swiss naturalist and Harvard professor Louis Agassiz joined the group in 1846 and made use of Bache's Coast Survey for marine biology observations. The 1850s brought two junior members to the Lazzaroni to complete the core. Astronomer Benjamin A. Gould brought his European credentials to the group including a Ph.D. from Göttingen and the patronage of Carl Friedrich Gauss. Chemist Oliver Wolcott Gibbs, Gould's companion in Europe, brought a New York connection and skill as a wielder of influence.[51]

The Lazzaroni wanted "self-rule for science, support without strings, the time and money to do research without having to account to laymen for its direction or consequences."[52] Self-rule for science was partially accomplished by getting themselves appointed to key scientific positions, Bache's superintendency of the Coast Survey and Henry's appointment at the Smithsonian, for example. The case of Wolcott Gibbs caused a notable battle. After the

---

[49]See Hubbell and Smith, 1992, p. 261

[50]Peirce signed at least one of his letters to Bache "your ever loving ϕ^{r}" and he referred to his wife as ϕ^{ras}. See Peirce to Bache, 27 March 1863, in Reingold, 1964, pp. 206–207.

[51]See Bruce, 1987.

[52]See Bruce, 1987, p. 225.

Lazzaroni failed to secure him a position at the University of Pennsylvania, they attempted to get him an appointment in chemistry at Columbia. They again failed after the chemist's candidacy became embroiled in issues of religious freedom and the New York newspapers entered the fray. In spite of the failure with Gibbs, however, the controversy appears to have eventually set Columbia on a positive course.[53]

For the "time and money to do research," they looked to private contributors who would give college faculties money for scientific equipment, and to the federal government for support of institutions like the Coast Survey. Bache managed to obtain $500,000 a year from the government for the Coast Survey, which was "undoubtedly the largest employer of mathematicians, astronomers, and physicists in antebellum America" and from whose helm Bache could "promote and stimulate scientific work in the United States."[54] Benjamin Peirce took over the Coast Survey after Bache's death in 1867. Surprisingly, he was an efficient administrator of the organization who "...maintained Bache's high standards of precision and finally realized Bache's old dream by commencing triangulation along the thirty-ninth parallel to connect the Atlantic and Pacific surveys, thus giving the Coast Survey a continental mission."[55]

For many years, "Bache had argued privately and publicly for an American equivalent of the French Academy, with its government subsidies for research, its publications, its spur to scientific work through public honor, its advice to government, and its perpetual secretary..." In 1862–1863, the idea of the National Academy of Sciences took shape among the Lazzaroni. Henry opposed the idea, partly on the grounds that the Academy might come under attack by those not included and that government support might lead to politicization. The bill forming the Academy passed and the group was organized amid a great deal of bickering. But the Academy was

---

[53] See Bruce, 1987, pp. 226–230.
[54] See Reingold, 1967, p. 152.
[55] See Bruce, 1987, p. 318.

not successful in securing government funding as planned. It would, however, survive to fulfill its role as a "touchstone of recognition" for American scientists.[56]

From its inception until Bache suffered a stroke in 1864, effectively ending the days of the Lazzaroni, the group pursued the advancement of professional competence in the sciences, meddled in practically every aspect of scientific life in the country and developed relationships within the government that led to federal funding for the sciences.[57] Their influence was not always impartial, and not always appreciated, and they frequently did not get along with each other. But historian of science Robert V. Bruce sums up the importance of their role by terming them the "architects of American science."[58]

Benjamin Peirce's contribution to the foundation for the transformation of mathematics at Harvard and the United States took three directions. For future generations, he set the example of an American research mathematician. Through his membership in the Lazzaroni, he embodied a spirit of scientific inquiry and activism that began to give form to an American scientific (including mathematical) infrastructure. And although he did not directly train a core of research mathematicians to follow him, he succeeded in passing his inspiration and vision to a next generation of educators. That next generation of educators would succeed in making his vision come true—they trained William Fogg Osgood, Maxime Bôcher and others who would be important members of the soon-to-be emerging American mathematical research community. Peirce passed the legacy of his vision to the next generation of educators, including his son James Mills Peirce, William Byerly, the recipient of Harvard's first Ph.D., and mathematical physicist Benjamin Osgood Peirce (a distant relative). It also included the next two

---
[56]See Bruce, 1987, p. 301–305

[57]See Bruce, 1987 for a detailed account of the trials and tribulations, successes and failures of the Lazzaroni.

[58]See Bruce, 1987, p. 217.

presidents of Harvard, Charles William Eliot, who held the post from 1869 to 1909 and Eliot's successor Abbott Lawrence Lowell.

The first period in American mathematics was coming to an end. At Harvard in particular, the hold of British traditions in mathematics teaching had loosened to admit the French analytic approach. A vision of American excellence in science and mathematics had blossomed and Benjamin Peirce had set the standard at Harvard, producing high quality research in mathematics and astronomy. J.J. Sylvester would soon arrive at the Johns Hopkins to inspire his core group of graduate students dedicated to mathematical research. Peirce's students were in position to usher in the period of emergence of a mathematical research community at Harvard, most notable among those students—Charles William Eliot.

2.2 Charles Eliot's Vision of American Higher Education and the Emergent Period of Research Mathematics in the United States

Victorian America was fascinated with new technology and innovations in the organization of society. It was during this period that the foundations of modern life were being put in place. Educational institutions in general were evolving into their current form. Historian of education Lawrence Cremin characterized the period 1876–1917 as that of the progressive impulse in education.[59] Americans were dissatisfied with the state of education at all levels. Educators were shocked at how little students learned from the rote recitation methods of the time and wanted to teach students to think and reason for themselves. This has a familiar ring to educators even into the third millennium.

[59] See Cremin, 1961.

Francis W. Parker was one of the pioneers of the progressive education movement. "Colonel" Parker spent two-and-a-half years in Europe (including Germany) observing the latest pedagogical ideas and returned to the United States with a desire to introduce pedagogical innovation in the American system. He became superintendent of the Quincy, Massachusetts schools in 1873 and developed the acclaimed "Quincy System" emphasizing observation, description and understanding. John Dewey, a later pioneer of the progressive movement in education, established the Laboratory School (associated with the University of Chicago) in 1896. The progressive movement, while primarily a force in primary and secondary education, also influenced and was influenced by those involved in higher education. Charles William Eliot, while representing an elitism that was not in line with some progressivist ideals, was a prominent and respected advocate of a new education. Daniel Coit Gilman's vision of the Johns Hopkins as an institution devoted to scientific methods of inquiry at the highest level can be viewed as another expression of the progressivist ideal. Gilman at Hopkins and William Rainey Harper at Chicago had the advantage of being able to start from the beginning in creating progressive new institutions. Surrounded by years of tradition, Eliot had to proceed more slowly at Harvard, but with no less conviction. It was in the midst of this fervor for reform in education at all levels that Osgood and his colleague Bôcher went through their formative years.

2.2.1 Eliot's vision

Charles William Eliot became president of Harvard College in 1869 at the age of 35. In his inaugural address, he began to set out a vision of what the American university might become:

> The endless controversies whether language, philosophy, mathematics, or science should supply the best mental

training, whether general education should be literary or chiefly scientific, have no practical lesson for us today.... This university recognizes no real antagonism between literature and science, and consents to no such narrow alternatives as mathematics or classics, science or metaphysics. We would have them all, and at their best.[60]

Eliot wanted Harvard to have the most effective methods of instruction available; he wanted languages taught systematically, science taught more inductively, mathematics and history taught more vividly and philosophy less dogmatically. He wanted written admission examinations, better pay for professors who were "the living sources of learning and enthusiasm" and greater flexibility for professors in their teaching.[61] He pointed to mathematics at Harvard (which was then under Peirce's leadership) as an example to be emulated:

If there be any subject which seems fixed and settled in its educational aspects, it is the mathematics; yet there is no department of the University which has been, during the last fifteen years, in such a state of vigorous experiment upon methods and appliances of teaching as the mathematical department. It would be well if the primary schools had as much faith in the possibility of improving their way of teaching multiplication.[62]

Eliot's vision was not without blind spots, however. He was not a champion of higher education for women, preferring to "maintain a cautious and expectant policy"[63] while the larger society debated the issue. That he did not expect a quick resolution to the issue was made clear when he continued:

---

[60] See Eliot, 1869a, p. lix.
[61] See Cremin, 1988, p. 379.
[62] See Eliot, 1869a, p. lxii.
[63] See Eliot, 1869a, p. lxx.

> The world knows next to nothing about the natural mental capacities of the female sex. Only after generations of civil freedom and social equality will it be possible to obtain the data necessary for an adequate discussion of woman's natural tendencies, tastes, and capabilities. Again, the [Harvard] Corporation do not find it necessary to entertain a confident opinion upon the fitness or unfitness of women for professional pursuits. It is not the business of the University to decide this mooted point.[64]

In an 1899 address, President M. Carey Thomas of Bryn Mawr chided Eliot for his views on education for women calling them a "dark spot of mediaevalism" on his "otherwise luminous intelligence."[65]

In the inaugural address, Eliot also considered his role as Harvard President in making his vision a reality. It was not to oversee details such as the purchase of doormats but to:

> ...watch and look before—watch, to seize opportunities to get money, to secure eminent teachers and scholars, and to influence public opinion towards the advancement of learning; and look before, to anticipate the due effect on the University of the fluctuations of public opinion on educational problems; of the progress of the institutions which feed the University; of the changing condition of the professions which the University supplies; of the rise of new professions; of the gradual alterations of social and religious habits in the community.[66]

He wanted the university to be uniquely American, so while it could in some respects resemble a European (German, in partic-

---

[64] See Eliot, 1869a, p. lxx.
[65] See Thomas, 1965, p. 142, in Cremin 1988, p. 383.
[66] See Eliot, 1869a, p. lxxvi.

2. Before Osgood and Bôcher

ular) model, he felt the need for the university to be responsive to American reality, observing that: "In this mobile nation the action and the reaction between the university and society at large are more sensitive and more rapid than in stiffer communities."[67] His inaugural address in 1869 was enthusiastically received and recognized as a "turning point in American higher education."[68]

2.2.2 The Roots of Eliot's vision for Harvard and its realization

Where did the ideas in Eliot's inaugural address spring from? What were his qualifications for the transformation of Harvard into a model American university? Eliot did not deliver his inaugural address as a stranger newly arrived at Harvard. His father was treasurer of the college from 1842 to 1853. Two of his aunts married Harvard professors. Eliot attended Harvard himself, graduating in 1853, becoming a tutor of mathematics and then assistant professor of mathematics and chemistry. But he left Harvard in 1863, disappointed at not having been selected for the Rumford professorship. He went to Europe to travel and study in French and German institutions of higher learning, working for a time at the laboratory of chemist Hermann Kolbe. He returned to the United States and in 1865 became professor of chemistry at the newly founded Massachusetts Institute of Technology where he remained until becoming President of Harvard in 1869.

Thomas Hill, Eliot's predecessor at Harvard, expressed some open-mindedness about educational innovation. He also wanted, on the other hand, to "more carefully segregate liberal education from the taint of vocationalism" and to ensure that intellectual training did not interfere with Harvard students being open to "simple and refining pleasures."[69] Eliot's appointment at Harvard was thus not

---
[67] See Eliot, 1869a, p. lxxvi.
[68] See Cremin, 1988, p. 380.
[69] See Veysey, 1965, p. 8.

without controversy. Unlike Hill, he was a known reformer, and he was critical of the Lawrence Scientific School. In an 1869 article for the *Atlantic Monthly* he reviewed then current attempts to change college curricula and warned against patterning the American university on European models. "The American university has not yet grown out of the soil...," wrote Eliot. "When the American university appears, it will not be a copy of foreign institutions, or a hot-bed plant, but the slow and natural outgrowth of American social and political habits, and an expression of the average aims and ambitions of the better educated classes. The American college is an institution without a parallel; the American University will be equally original."[70]

The Harvard he was speaking to in his inaugural address and thinking about as he wrote his *Atlantic Monthly* article was small and still local in character. Harvard College itself had a student population of about 500 undergraduates and a faculty of 23. The curriculum included Latin, Greek, German, French, mathematics, some philosophy, history, physics, chemistry and a few electives in sciences and modern languages. The law school granted a degree to any student in residence eighteen months. In the medical school, two terms, an apprenticeship and a ten-minute oral exam earned a student an M.D. degree. The Divinity School "for all intents and purposes did not grant degrees," while the Lawrence Scientific School had an eminent faculty but "pitifully low standards for entry and exit."[71]

Largely through Eliot's work and influence, Harvard became larger in size, cosmopolitan in character, international in stature. Its undergraduate offerings broadened. The Lawrence Scientific School was basically incorporated into the college. The Graduate Department was established in 1872, offering master's and doctor's degrees. The professional schools were placed on a graduate level,

---

[70] See Eliot, 1869b, p. 216.
[71] See Cremin, 1988, p. 382.

and high quality faculty members were recruited. By 1894, Eliot's 25th year in office, Harvard was a shining model of his vision of the new American university.[72]

William Fogg Osgood and Maxime Bôcher received their undergraduate education at Harvard and began their faculty careers during Eliot's presidency. They were part of Eliot's product, a realization of his vision of Harvard as the new American university. How did Eliot's vision manifest itself in the mathematics department at Harvard? What were the elements that led to Osgood and Bôcher being among the members of the emerging mathematical research community? An examination of four key faculty members helps answer those questions. These faculty members formed the bridge between Benjamin Peirce on one side of the transformation, and Osgood and Bôcher on the other, and they were the implementers of Eliot's vision as it pertained to mathematics.

2.3 The Bridge between Peirce and his Mathematical Successors, Osgood and Bôcher

Benjamin Peirce encouraged research interests on the part of faculty members that came after him, among them his son James Mills Peirce, William Byerly, and B.O. Peirce.[73] Byerly and James Mills Peirce came to be more known for their strength in teaching.[74] B.O. Peirce became active in research in mathematical physics and his work in astronomy was a factor in attracting two experts in astronomy to the Cambridge community: Simon Newcomb, a graduate of the Lawrence Scientific School, and G.W. Hill. Newcomb and Hill

---

[72] See Cremin, 1988, p. 382.

[73] Benjamin Peirce's most famous son was Charles Saunders Peirce, who did not join the faculty at Harvard. He was a brilliant yet temperamentally difficult logician who lectured at the Johns Hopkins for five years and did some work for Harvard's observatory, but did not have a successful academic career.

[74] See Parshall and Rowe, 1994, p. 20.

were the third and fourth presidents of the American Mathematical Society.[75]

James Mills Peirce (1833–1906), who received his A.B. from Harvard in 1853, carried the burden of a reputation for not being as brilliant as his famous father. But, J.L. Coolidge noted that J.M. Peirce played the dual role of a person interested in teaching mathematics and, while not being a great scholar, someone who appreciated scholarship in others and fostered graduate mathematics at Harvard in its early years. In particular, he was Eliot's collaborator in promoting graduate education and he took charge as secretary of the academic council of the new graduate department.[76] Coolidge further credited him with "fostering an interest in the theory of functions of a complex variable, a subject which subsequently became the very backbone of the instruction in higher mathematics,"[77] and the general arena of Osgood's main research interest.

Coolidge also credited this younger Peirce with starting a tradition of senior and junior faculty members sharing equally in the most sought-after teaching assignments. The senior faculty members could no longer take all the most advanced courses for themselves, leaving the "crumbs" to the junior faculty. This gave junior faculty members opportunities to show their abilities, and put freshmen and sophomores in contact with the sometimes more eminent senior faculty members. This became a popular policy at Harvard and, along with J.M. Peirce's skill in teaching; it dispelled some of the myth left in Benjamin Peirce's wake that a great scholar must also be a "lamentable" teacher.[78]

William Elwood Byerly (1850–1934) received his A.B. in 1871 and went on to receive the first Doctor of Philosophy degree ever awarded at Harvard. His 1873 thesis, "The Heat of the Sun," at-

---

[75] See Birkhoff, 1989, p. 7.
[76] See Birkhoff, 1989, p. 12.
[77] See Coolidge, 1930, p. 250.
[78] See Coolidge, 1930, p. 253.

2. Before Osgood and Bôcher

tempted a calculation of the total energy of the sun, using calculus and elementary thermodynamics.[79] He was known as a successful teacher, instilling a love for mathematics in his students and publishing the best American texts of the time in differential and integral calculus.[80] Coolidge faults Byerly for loving his classes more than his subject and for not maintaining an interest in scholarship.[81] Byerly and B.O. Peirce, in 1883–1884, developed a course in mathematical physics which Garrett Birkhoff called "truly remarkable" noting that it still continued to be taught at Harvard with suitable revisions.[82]

Benjamin Osgood Peirce (1855–1913) was known as more of a scholar than his colleagues. Most of his published work was in physics, but he also produced *A Short Table of Integrals*[83] that was widely used (supplementing Byerly's calculus text). Coolidge remembered him for introducing the take-home examination (called the long paper) to supplement the mid-term or final exam.[84] One of his take-home exams contained 50 problems which, with parts, amounted to 133 distinct questions! B.O. Peirce succeeded Joseph Lovering, becoming the Hollis Professor of Mathematics and Natural Philosophy in 1888, and served as President of the American Physical Society in 1913. Osgood's interest in potential theory may be partly the result of B.O. Peirce's own interest in the subject and of his textbook, *Theory of the Newtonian Potential Function*[85], for which Osgood later offered suggestions as a colleague when Peirce revised the text.

Garrett Birkhoff's assessment is that these three men "regarded their profession as that of *teaching* reasonably advanced mathemat-

---

[79] See Birkhoff, Garrett, 1989, p. 10.
[80] Among these, Garrett Birkhoff mentions Byerly's *Differential Calculus* (1879), *Integral Calculus* (1881) and an adaptation of Chauvenet's *Geometry* (1887) which were widely used in other American colleges (Birkhoff, 1989, p. 14).
[81] See Coolidge, 1930, p. 250.
[82] See Birkhoff, 1989, p. 14.
[83] See Peirce, 1929.
[84] See Coolidge, 1930, p 250.
[85] See Peirce, 1888.

ics in an understandable way."[86] They counted a number of influential men among their students, including Arthur Gordon Webster, another president of the American Physical Society, Osgood and Bôcher, and Frank Nelson Cole.

Frank Nelson Cole (1861–1926), a few years senior to Osgood and Bôcher, was a product of the department led by the two Peirces and Byerly. Cole preceded Osgood and Bôcher to Germany and his success both there and upon his return to join the Harvard faculty paved the way for them. Cole excited Osgood and Bôcher with his tales of study at Leipzig and with the mathematics he taught. Cole studied with Felix Klein at Leipzig in 1884. Emulating Cole, Osgood and Bôcher would also study with Felix Klein, but after the latter's move to Göttingen. Osgood attended Cole's lectures in 1885 at Harvard and four decades later Osgood recalled:

> [Cole] had just returned from Germany and was aglow with the enthusiasm which Felix Klein inspired in his students. Cole was not the first to give formal lectures at Harvard on the theory of functions of a complex variable, Professor James Mills Peirce having lectured on the subject in the seventies. That presentation was, however, solely from the Cauchy standpoint... [whereas] Cole brought home with him the geometric treatment [of complex functions] which Klein had given in his noted Leipsic [sic] lectures in the winter of 1881–82.

Osgood continued, offering the opinion that:

> [J.M. Peirce's lectures] stood as the Old over and against the New and of the latter Cole was the apostle. The students felt that he had seen a great light. Nearly all the members of the Department attended his lectures. It

---
[86]See Birkhoff, 1989, p. 14.

was the beginning of a new era in graduate education at Harvard, and mathematics has been taught here in that spirit ever since. (Osgood, in Fiske, pp. 773–774, as appearing in Parshall and Rowe, 1994, p. 196)

Osgood's elegy of Cole aside, Cole had only a short and difficult career at Harvard. However, his early heralding of a new era in graduate education at Harvard was an indicator that the transformation of Harvard mathematics was indeed happening. Cole's studies with Felix Klein in Germany gave a clear signal to Osgood and Bôcher that an American could indeed be successful at the highest levels in mathematics.

William Fogg Osgood and Maxime Bôcher were in the mathematical generation that followed J.M. Peirce, William Byerly and B.O. Peirce at Harvard. Harvard and German-educated, renowned scholars and well-connected to the mathematical world, Osgood and Bôcher became intricately woven into the transformation in mathematics at Harvard around the turn of the century, a transformation with roots in the climate created by the leadership of Byerly, Eliot, Cole and the various members of the Peirce clan.

The Harvard of the early 1890s when Osgood and Bôcher joined the faculty was thus a vastly different place from the small colonial college where Isaac Greenwood taught mathematics. Osgood and Bôcher did not teach all the sciences like their predecessor John Winthrop did, and therefore they were better able to specialize in their mathematical fields. Harvard offered its students a stronger mathematical education, including graduate studies that prepared them to be able to study abroad with mathematicians like Felix Klein. James Mills Peirce, William Byerly and B.O. Peirce had provided that stronger education to Osgood and Bôcher. Under the influences of Peirce and Eliot, achievements in pure research had become a desirable goal for faculty members. The mathematics department that welcomed Osgood and Bôcher back from Germany, inspired by Cole, was ready to join a larger mathematical world. Osgood and Bôcher quickly became Harvard's most distinguished representatives of the newly emerging American mathematical research community as they began their careers at Harvard in the early 1890s. They were, of course, part of a larger trend in the country. In particular, E.H. Moore and L.E. Dickson represented the fledgling American community at the University of Chicago.

The next three chapters will focus in more narrowly on three of Osgood's most important research results published in 1897, 1900 and 1903 in order to show how one representative of the new mathematical research community helped put the United States on the mathematical map of the world. In doing so, Osgood would be instrumental in completing the transformation of mathematics at Harvard. On the national level, his research results represent a part of the bridge between the period of emergence of the mathematical research community and the period of consolidation and growth.

Chapter 3

Non-Uniform Convergence and the Integration of Series Term by Term

One of Osgood's first major research papers, "Non-Uniform Convergence and the Integration of Series Term by Term" was presented to the American Mathematical Society (AMS) on August 31, 1896. It appeared in the *American Journal of Mathematics* in 1897.[1] This was Osgood's primary offering to resolving the question of the conditions under which, given a series,

$$u_1(x) + u_2(x) + \cdots + u_n(x) + \cdots,$$

the terms of which are continuous functions of a real variable x with partial sums,

$$s_n(x) = u_1(x) + u_2(x) + \cdots + u_n(x),$$

the integral of the limit of the partial sums is equal to the limit of the integrals of the partial sums. That is when is it true that

$$\int_{x_0}^{x_1} \lim_{n \to \infty} s_n(x)\, dx = \lim_{n \to \infty} \int_{x_0}^{x_1} s_n(x)\, dx?$$

---
[1] See Osgood, 1897a.

Henri Lebesgue's dominated convergence theorem would answer the question definitively in 1908.[2] Osgood called a question like the one above a double limit problem since it involves examining the conditions under which the order of taking two limits can be inverted.[3] He continued to demonstrate an interest in questions of "double limits" in his textbooks and elementary writings published after 1897, but he turned primarily to other subjects for his research.

Not only was the 1897 paper one of Osgood's first major research papers, it was also among the first papers by an American to receive international attention as part of the mainstream mathematical dialogue in Europe. It helped put the United States on the mathematical map of the world by attracting the interest of some of the best European mathematicians. It was also a step in the process of establishing an American research tradition, providing a potential model for emerging American mathematicians to emulate.

Osgood divided his paper into two parts. In Part I on "Non-Uniform Convergence," the results of greatest interest are actually topological theorems related to the classification of certain sets as being of the first or second category, as René Baire later termed them, in particular Osgood's independent version for the real line of what came to be known as Baire's theorem. A set is said to be of first Baire category if it can be expressed as a union of countably many nowhere dense sets. A set which cannot be so expressed is said to be of second category. A set L contained in L_1 is dense in L_1 if for every x_1 in L_1 and $\varepsilon > 0$ there is a point $x \in L$ such that

$$0 < |x - x_1| < \varepsilon.$$

---

[2] See Lebesgue, 1908. Lebesgue's dominated convergence theorem for non-negative functions appears in *Measure and Integral* by Wheeden and Zygmund as follows: Let $\{f_k\}$ be a sequence of nonnegative measurable functions on E such that $f_k \to f$ almost everywhere with respect to Lebesgue measure in E. If there exists a measurable function φ such that $f_k \leq \varphi$ for all k and if $\int_E \varphi$ is finite, then $\int_E f_k = \int_E f$. A property holds true *almost everywhere* if it is true except in a set of measure zero.

[3] This terminology was later mentioned by Émile Borel in his 1905 *Leçons sur les fonctions de variables réelles*.

3. Term by Term Integration

So L is *dense* in L_1 if every point in L_1 is a limit point of L. For example, the set of rational numbers is dense in the set of real numbers. A set L is *nowhere dense* in L_1, if there is no neighborhood of L_1 in which L is dense. The set of integers, for example, is nowhere dense in the set of real numbers, as is the set

$$\left\{1, \frac{1}{2}, \frac{1}{3}, \ldots\right\}.$$

In Part II on "Term by Term Integration" he proved his main result: If a sequence of continuous functions is bounded by a fixed constant and converges pointwise to a continuous function then the sequence can be integrated term by term, i.e.

$$\int_{x_0}^{x_1} \lim_{n \to \infty} s_n(x)\, dx = \lim_{n \to \infty} \int_{x_0}^{x_1} s_n(x)\, dx.$$

If a sequence of continuous functions converges uniformly on an interval, it is not difficult to prove convergence under the integral sign.[4] Thus it is the case of non-uniform convergence that causes the trouble. In Part I Osgood focused attention on the set of points at which the convergence is particularly non-uniform and proved his Fundamental Theorem; namely, that this set is nowhere dense. He then used this result in Part II to prove the term by term integration theorem.

It is interesting to note that, while the term by term integration theorem was Osgood's main interest and the focus of most of the attention at the time, it is the topological work in Part I that remained of enduring interest to mathematicians of future generations. Lebesgue's dominated convergence theorem moved Osgood's main theorem into a category of historical rather than current mathematical significance.

---

[4] A sequence of functions converges uniformly to a function f on an interval E if for every $\varepsilon > 0$ there is an integer N such that $|f_n(x) - f(x)| < \varepsilon$ for $n \geq N$ and for all x in E.

3.1 Some Historical Context—A Flurry of Measures and Integrals

The advent of point set theory led, in the 1880s, to the need to assign "measures" of length, area or volume to the various strange sets that were under examination. The word *measure* has both an intuitive, colloquial meaning—the board measures 8 cm long—and has developed a more precise mathematical meaning in the field of measure theory, a field which is partly the result of efforts described in this chapter. A function μ is a measure on a set L if $0 \leq \mu(L) \leq \infty$ and any countable collection of disjoint subsets of L, $\{L_k\}$, has the countable additivity property,

$$\mu\left(\bigcup L_k\right) = \sum \mu(L_k).$$

Note that a measure in this mathematical sense embodies the characteristics that a familiar measure, such as a length or an area, should have—it is a non-negative number or infinite and the measure of a collection of disjoint pieces adds up to the measure of the whole form.

According to Florian Cajori, the earliest precursors of the measures of measure theory were due to Hermann Hankel and Axel Harnack around 1882.[5] Georg Cantor offered his suggestions in 1884. Harnack published a new paper on the subject in the *Mathematische Annalen* in 1885, giving credit to Cantor for some of the results.[6] Harnack focused on what he called the content of a set, specifically a subset of the real numbers, $\mathbb{R}$. The idea of content was meant to be a more general concept of length. The notion of length does not apply to most subsets of the real line, but Harnack and others needed to be able to measure these subsets. The use of various definitions of content was a precursor to modern measure, and different mathematicians defined content differently. A content, for

---
[5] See Cajori, 1924, pp. 404–405.
[6] See Harnack, 1885.

3. Term by Term Integration

example, would not have the countable additivity property. Precise, formal definitions of content may not have been of much interest initially, since use of a content was sometimes seen as a process to reach an end-result, rather than an object of study in its own right.

In this 1885 paper, Harnack defined the content of a subset of the real line as the infimum of the lengths of coverings of the set by a *finite* number of non-overlapping closed intervals.[7] He also provided the following procedure for calculating this content. Suppose that the set G to be measured lies in an interval L of length λ. Disjoint open intervals are removed from $L - G$ in stages. The requirement is that after stage n has been accomplished, what is left of $L - G$ does not contain any open intervals of length greater than or equal to $1/n$. Let σ be the sum, possibly infinite, of the lengths of all the intervals removed. Then the content of L is the difference $\lambda - \sigma$. Because Harnack required that G be covered by a finite number of intervals, the content of the rationals in the unit interval, indeed of any dense subset, is 1.

Use of the word *measure* is difficult to avoid in a discussion of the mathematical notion of content. In the remainder of this chapter, the word *measure* is used in an informal sense to simply refer to a "means of measurement" of a set. Harnack's *content* in this usage is an exterior measure.[8] As a procedure for calculating that exterior measure, he used the interior measure of the complement of G described above. However, his interior measure used a countably *infinite* number of intervals. A process involving an infinite number of intervals was, however, not yet considered a candidate for reaching a good notion of measure. For Harnack and others, including Osgood whose procedure was similar to Harnack's, it was just a procedure to reach a desired result. Harnack did not extend his notion of content to the obvious choice of coverings by a finite number of

---

[7] A *covering* or *cover* of a set G contained in E is a family of subsets of E, E_k, such that G is contained in their union, $\bigcup E_k$. A covering can itself contain subcovers.

[8] Think of an exterior measure as one obtained by covering the interval, while an interior measure is obtained by filling the interval, i.e., the difference between an upper and lower Riemann-Darboux sum.

suitable non-overlapping closed rectangles. Instead he adapted the procedure described above to remove n-dimensional spheres from an n-dimensional interval containing the set G to be measured.

The notions of content evolved from the concept of area as extended to point sets, and areas were traditionally defined by means of Riemann integrals. Precisely what constituted the new link between content and the Riemann integral had also to be determined. The work of Giuseppe Peano provided the bridge. Making more specific some ideas he had expressed in 1883[9], he treated questions of measure and integral in his 1887 *Applicazioni geometriche del calcolo infinitesimale*.[10] While still insisting on finite coverings, Peano did define both an interior and an exterior area. It thus made sense to talk about *the* area of a region if both the interior and the exterior area coincided. For a non-negative function f on an interval, Peano noted that the upper Riemann integral corresponds to the exterior area of the region under the graph of f while the lower Riemann integral corresponds to the interior area. (In fact, this is what came to be known as the Darboux version of the Riemann integral, defined via converging upper and lower sums.) Thus, f is integrable on that interval if and only if the region under the graph is measurable, meaning that the interior and the exterior areas are equal. Peano handled corresponding definitions for one and three dimensions as separate cases.

Five years later, Camille Jordan defined his notion of measurability more generally, i.e. for n-dimensional bounded sets, but via inner and outer content in a manner essentially similar to that of Peano. A set was said to be measurable if its outer content was equal to its inner content. Jordan was among the first to recognize the importance of considering the finite additivity of measure. Regarding the theory of integration as developed by Bernhard Riemann and Gaston Darboux, Jordan wrote in 1892 that:

---

[9] See Peano, 1883.
[10] See Peano, 1887, Geometric applications of the infinitesimal calculus.

3. Term by Term Integration

The influence of the nature of the domain does not appear to have been studied with the same care [as the role of the function in the integral]. All the demonstrations rest upon this double postulatum: that each domain E has a determinate extension; and that, if it is decomposed into several parts $E_1, E_2, \ldots$, the sum of the extensions of these parts is equal to the total extension of E. But these propositions are far from being evident if full generality is allowed to the concept of the domain.[11]

In response to this concern, Jordan proved that if a set E is the disjoint union of a finite number of measurable sets E_p then E is measurable and its measure is the sum of the measures of the E_p. His ideas on measure and integration soon became widely used and accepted, appearing in 1893 in his popular *Cours d'Analyse de l'École Polytechnique*.[12]

Harnack, Peano and Jordan all defined content using finite coverings. Borel broke with this tradition when in 1898 he made the breakthrough of measuring a set via a covering by a *countably infinite* number of intervals, and introducing the notion of countable additivity. Henri Lebesgue (1875–1941) generalized further the notion of measure and applied it to the theory of integration in his famous doctoral thesis, "Intégrale, longueur, aire," published in 1902.[13]

Where exactly does Osgood's 1897 paper fit into this historical account? Osgood was familiar with Harnack's 1885 work as well as with the work by Cantor; he mentioned both in this paper.[14] Moreover, he used Harnack's content not to define an integral but

---

[11] See Jordan, 1892, pp. 69–70 as translated in Hawkins, 1970, pp. 93–94. Much of the discussion of Peano and Jordan measure is based on this source.

[12] See Jordan, 1893, Analysis Textbook of the Polytechnic School.

[13] See Lebesgue, 1902, Integral, length and area. For a full discussion of the characteristics of the measures as defined by Peano, Jordan, Borel and Lebesgue, see Hawkins, 1970.

[14] In particular, Osgood was familiar with Cantor's "Fondements d'une théorie générale des ensembles," his "Sur divers théorèmes de la théorie des ensembles," and other papers translated into French and reprinted in *Acta Mathematica*, vol. 2, 1883.

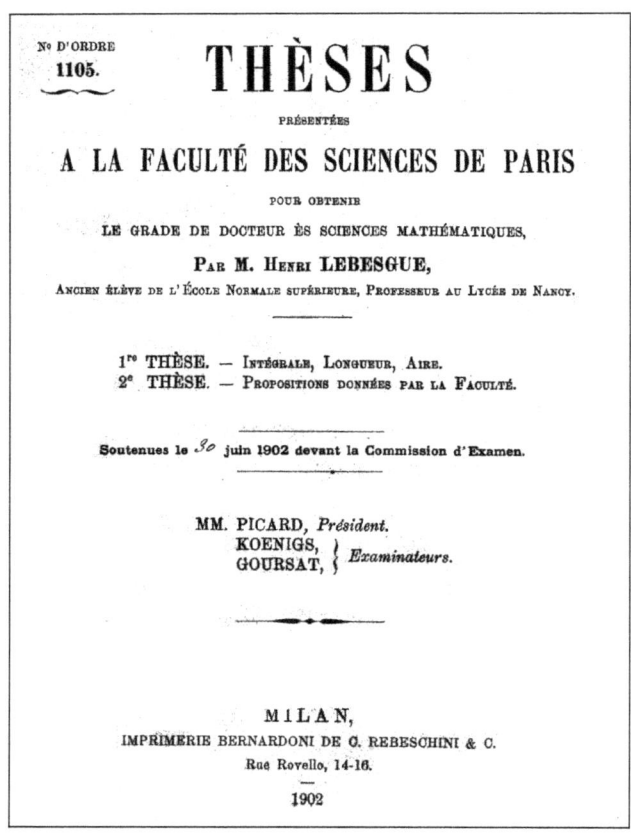

Figure 3.1: Title Page of Lebesgue's Thesis

instead as a step in the process of calculating a Riemann-Darboux integral (converging upper and lower sums). It is unclear whether or not he was aware of Peano's 1887 work at this time, although he was well aware of it by 1903. He also gave no indication of being aware of Jordan's work.

Osgood's paper came at an interesting juncture, five years before Lebesgue published the 1902 dissertation that was to influence the field so profoundly. In general, the Lebesgue integral was then somewhat slow, at least as seen by more modern standards, to be accepted by the American mathematical community. It was also slow

3. Term by Term Integration

to be accepted in some European mathematical circles. Lebesgue's dissertation was only accepted over opposition, and his ideas continued to be criticized for some time to come. Perhaps a factor in the opposition and slow acceptance was the unexpectedness of the new integral. Kenneth May wrote that "Lebesgue's definition of the integral was not 'in the air' and might have remained undiscovered for another fifty years," and Lebesgue himself expressed doubts about the value of his work.[15] But this work gained acclaim and eventually wide acceptance. In 1917, Gilbert Ames Bliss gave an introductory lecture on the subject of Lebesgue integrals at an AMS symposium in Chicago.[16] Bliss wrote: "It would be impossible to have selected a subject for that occasion more characteristic of present mathematical tendencies." In describing the mood of the time, Bliss continued:

> In the field of integration the classical integral of Riemann, perfected by Darboux, was such a convenient and perfect instrument that it impressed itself for a long time upon the mathematical public as being something unique and final. The advent of the integrals of Stieltjes and Lebesgue has shaken the complacency of mathematicians in this respect, and, with the theory of linear integral equations, has given the signal for a reexamination and extension of many of the types of processes which Volterra calls passing from the finite to the infinite.[17]

In 1917, Émile Picard expressed a similar sentiment when he wrote: "Riemann seemed to have investigated as deeply as possible the idea of the definite integral. Lebesgue showed that this was not at all the case."[18] The Lebesgue integral did not immediately become a widely used workhorse integral among mathematicians in

---

[15] See May, 1966, p. 2, with reference to Denjoy, 1949, p. 578.
[16] See Bliss, 1917.
[17] See Bliss, 1917, p. 1.
[18] See Picard, 1917, as translated and discussed in May, 1966, p. 2.

the United States; its adoption for both research and teaching purposes was gradual. Maxime Bôcher's student G.C. Evans, however, did teach the Lebesgue integral to undergraduates at Rice Institute as early as 1914.[19]

At the time Bliss wrote, the dust had not yet settled on the battle of the integrals. In addition to the integrals of Lebesgue and T.-J. Stieltjes, there were also integrals devised by William H. Young, James Pierpont, Johann Radon, Maurice Fréchet, E.H. Moore and others, some of which are still in use. Young was the husband of Grace Chisholm Young. Much of their work was done jointly but published under his name only. Their 1906 book (republished in 1972) "The Theory of Sets of Points" was an early systematic presentation of set theory.

Efforts were made to show that certain integrals could be reduced to another integral, the Lebesgue integral being the primary target. Bliss felt it unfortunate that this was interpreted by some as an effort to oust the Riemann integral from its place of honor and to replace it with the Lebesgue integral. To counter this interpretation, Bliss cited an article in which Edward Burr Van Vleck "remarked that a Lebesgue integral is expressible as one of Stieltjes by a transformation much simpler than that used by Lebesgue for the opposite purpose, and the Stieltjes integral so obtained is readily expressible in terms of a Riemann integral."[20] This assertion needs some qualification. There are, as Bliss noted, Lebesgue integrable functions which are not Riemann integrable, the Dirichlet function in [0, 1] for example.[21] However, Bliss shows in section six of his paper, that Van Vleck's remark is valid in the case of Jordan measurable sets, a subset of Lebesgue measurable sets.

---

[19] See May, 1966, p. 4.
[20] See Van Vleck, 1917, p. 327, as quoted in Bliss, 1917, p. 2.
[21] The *Dirichlet function* is defined by $f(x) = 1$ if x is rational and $f(x) = 0$ if x is irrational.

Van Vleck appears to have believed that eventually one of the flurry of integrals would come to dominate the field, and he did not consider Lebesgue's integral the top candidate. Bliss, a colleague of E.H. Moore at the University of Chicago, thought that a more "general theory of the type of those of Fréchet and Moore" would become the standard form. One reason the Lebesgue integral became so prominent is its utility in functional analysis, i.e. Hilbert and Banach spaces.

The answer to the question of where Osgood's paper fits into the historical account can be summarized, therefore, in two ways. In terms of the development of measure, it followed especially closely the ideas of Harnack, it did not make use of the ideas of Jordan and Peano which had already been published, and it was a precursor to the work of Borel and Lebesgue which was soon to come. In terms of integration, Osgood employed the obvious integral for the time—the Riemann-Darboux integral. As it was, he did not have to think about which integral to use; he did not have to deal with the proliferation of integrals that would arise over the following few years.

3.2 Part I of the 1887 Paper—Non-Uniform Convergence

In an earlier 1896 paper, Osgood considered a geometric method for the treatment of *uniform* convergence,[22] and this 1897 paper examined the obvious next case, building on Harnack's 1885 work. In a finite closed interval $[a, b]$, let

$$f(x) = u_1(x) + u_2(x) + \cdots + u_n(x) + \cdots$$

where the u_i and f are continuous functions in x and the partial sums $s_n(x)$ converge pointwise to $f(x)$. In Part I of his paper,

---
[22] See Osgood, 1896.

Osgood focused particularly on the cases where convergence is *not* uniform and he gave several carefully selected examples in order to illustrate the conditions of his theorem. Osgood's first example has closed form partial sums

$$s_n(x) = \frac{n^2 x}{1 + n^3 x^2},$$

and his second has

$$s_n(x) = \frac{nx}{1 + n^2 x^2},$$

both on $[0, 1]$ with $f(x) \equiv 0$.

In one of Osgood's examples, the peaks of the partial sums $s_n(x)$ are unbounded as n goes to infinity, while in another the peaks have a maximum value. He called points where the peaks are unbounded X-points. More precisely, α is an X-point if given any neighborhood, T, of α and any real number $A > 0$ there are infinitely many values of n for which there exists x in T such that $s_n(x) > A$. Osgood intended to show that if $x_0 < x_1$ are any two points in an interval $[a, b]$ free of X-points then

$$\int_{x_0}^{x_1} \lim_{n \to \infty} s_n(x) dx = \lim_{n \to \infty} \int_{x_0}^{x_1} s_n(x) dx.$$

Otherwise stated, the series $\sum_{n=1}^{\infty} u_n(x)$ is integrable term by term.

To illustrate the usefulness of this result, Osgood exhibited a function whose convergence is non-uniform in every sub-interval of $[0, 1]$, but which by virtue of having no X-points can be integrated term by term. Let

$$\phi_k(x) = \frac{n \sin^2 k\pi x}{1 + n^2 \sin^4 k\pi x}$$

and

$$s_n(x) = \sum_{i=1}^{\infty} \frac{1}{i!} \phi_{i!}(x).$$

Then $f(x) = 0$ and $s_n(x)$ has peaks in the neighborhood of any rational x_0.

The assumption that the interval is free of X-points implies that there is a uniform bound on the partial sums; Osgood proved this by means of a Bolzano-Weierstrass argument. A more familiar argument might be based on the fact that if α is not an X-point, then we may to it assign a suitable neighborhood. These neighborhoods form an open cover which, by compactness, has a finite subcover. Osgood typically used the Bolzano-Weierstrass type argument in situations where the Heine-Borel property[23] could be used. This is to be expected since, although Heine used such an argument implicitly in the 1870s, it was not until 1895, not long before Osgood's paper, that Borel formally stated and proved it.[24]

3.2.1 The Fundamental Theorem

After Osgood provided the examples and context described above, he embarked upon the main result of Part I—his Fundamental Theorem. Let
$$s_n(x) = u_1(x) + u_2(x) + \cdots + u_n(x),$$
where the $u_i(x)$ are single-valued, real functions continuous in an interval $L = [a, b]$. Moreover, suppose $s_n(x)$ converges pointwise to a continuous function $f(x)$. Define
$$S_n(x) = s_n(x) - f(x),$$
so that
$$\lim_{n \to \infty} S_n(x) = 0.$$

---

[23] These arguments are closely related. Bolzano-Weirstrass Theorem: A bounded sequence of real numbers has a convergent subsequence. Heine-Borel Theorem: A subspace of the real numbers is compact if and only if it is closed and bounded, or equivalently if and only if every sequence has a convergent subsequence. A set is *compact* if and only if every open cover has a finite subcover.

[24] See Katz, 1993, pp. 735–736 for additional details.

Osgood fixed a positive quantity A. Then there are two types of points: those for which there exists $\delta > 0$ and a positive integer N such that for all $n \geq N$, if $|x - x_0| < \delta$ then $|S_n(x)| < A$, and those for which for all $\delta > 0$ and all positive integers N, there exists an $n \geq N$ and an x such that $|x - x_0| < \delta$ and $|S_n(x)| > A$. Osgood called these latter points γ-*points*. His Fundamental Theorem stated that the set G of γ-points for a given A is closed and nowhere dense in L.[25] The X-points mentioned above are those with infinite peaks and can be viewed as the intersection over all choices of A of the sets of γ-points.

Osgood used his Fundamental Theorem to prove his convergence result in Part II, but not before continuing on in Part I to examine some further issues related to the Fundamental Theorem. This was partly in response to a paper by Paul du Bois Reymond who believed he had exhibited a function whose X-points were dense in an interval and thus, by the nature of X-points, filled the interval.[26] So Osgood carried his result further to show that the X-points are nowhere dense in L.

The relationship of Osgood's discussion to what is now known as Baire's theorem is of considerable interest. The usual setting for Baire's theorem later came to be a complete metric space. From this point of view, the theorem states that a complete metric space cannot be represented as the union of a countable number of nowhere dense sets, or equivalently, that the complement of a union of a countable number of nowhere dense sets in a complete metric space is everywhere dense in that space. Baire proved his version of the theorem in 1899 but Osgood is credited for a first proof, for the special case of the real line, of what came to be known as Baire's theorem or the Baire category theorem.[27] To examine Osgood's result, let

$$\xi = \{x : \exists A \text{ for which } x \text{ is a } \gamma\text{-point for that } A\},$$

---

[25] The inequality $|S_n(x)| < A$ on page 161 of his paper should be replaced by $|S_n(x)| \leq A$
[26] Osgood gave as reference the *Sitz.-Ber. d. Berliner Akad.*, 1886, page 359.
[27] See for example Lukeš, Malý and Zajíček, 1986, p. 139.

i.e. ξ is a set of peak points. Taking A to be $1/n$ where n ranges over all positive integers, the set ξ can be written as a union of countably many closed nowhere dense sets. Let ζ be the complement of ξ relative to L. Thus

$$\zeta = \{x : \nexists A \text{ for which } x \text{ is a } \gamma\text{-point for that A}\},$$

i.e. ζ is the set of non peak (or null peak) points. Using the same method as in his Fundamental Theorem, Osgood showed that every subinterval of L contains a point of ζ so that ζ is at least dense in L. Thus, the set ξ is not the entire interval L.

Osgood did not use the fact that the sets are closed in the proof of this modification of his Fundamental Theorem. This is of interest since it is not necessary for the usual Baire's theorem that the sets be closed. Osgood's hypothesis, found throughout his paper, that sets be closed and nowhere dense is discussed later in this chapter.

Osgood realized the importance of the general properties of the sets for which he provided these results. He wrote that sets such as ξ belonged "to a class of sets of points of so much importance, ...that I will define this class independently of the considerations that have led to it"[28] and he provided a formulation of these properties. Osgood's formulation amounted to a definition of sets of Baire first category in the real line, except that in considering a union of nowhere dense sets he required these sets to be closed. But his proofs just mentioned did not use this requirement that the sets be closed. The sets he was interested in throughout the paper satisfied what he called *Conditions P*, i.e., they were closed and nowhere dense, and he did not examine which of the two conditions he used for each particular proposition.

Although Osgood did not need to elaborate further to prove his main result, he continued his exploration of this topic by proving that the points of ζ are uncountable in every subinterval of L. If

---

[28] See Osgood, 1987a, p. 171.

these points were not uncountable, ζ could be written as the countable union of singleton sets in some subinterval of L, say L_1. Then, in modern terms, the complete metric space

$$L_1 = (\xi \cap L_1) \cup (\zeta \cap L_1)$$

could be written as a countable union of closed, nowhere dense sets, a contradiction.[29]

The main purpose of Osgood's paper was to prove a result about term by term integration. In Part I, he examined cases in which partial sums do not converge uniformly to a limit function and he wrote at some length about properties of certain closed, nowhere dense sets. What does Osgood's emphasis on closed, nowhere dense sets have to do with his integration theorem? The second major result that Osgood used in his proof of the theorem is a convergence lemma, the statement of which will begin to shed some light on this question.

3.3 Part II of the 1887 Paper—Term by Term Integration

A result critical to Osgood's proof of the integration theorem is a convergence lemma, which is of standard type except for the fact that he required the limit set to be both closed and nowhere dense.[30] Osgood supposed G to be any set of points that is closed, nowhere dense and the union of a chain of subsets

$$G_1 \subseteq G_2 \subseteq \cdots \subseteq G_i \subseteq \cdots$$

Then, $\lim_{i \to \infty} \mu_H(G_i) = \mu_H(G)$. Here, Osgood was using Harnack's content.[31] This restricted him to the use of finite coverings. This

---

[29] See Osgood, 1897a, p. 173.

[30] See Osgood, 1897a, pp. 178–179.

[31] The symbol μ will be used for content and measure with the subscript indicating whose content or measure; μ_H for Harnack, μ_L for Lebesque, etc. Interior and exterior content or measure will be denoted μ^{int} and μ^{ext} respectively.

3. Term by Term Integration

was the first example of a lemma of this type, and it almost certainly influenced Lebesgue who used an analogous but simpler idea to prove his later theorem on term by term integration.[32]

Osgood's proof of the lemma can be outlined as follows. It is true that $\mu_H(G_j) \leq \mu_H(G_k)$ whenever $k > j$ and for any i, $\mu_H(G_i) \leq \mu_H(G)$. Thus $\lim_{i \to \infty} \mu_H(G_i)$ exists, and $\lim_{i \to \infty} \mu_H(G_i) \leq \mu_H(G)$.

For the opposite inequality, let $\delta > 0$ be given. Choose a decreasing sequence of δ_i such that $\sum_{i=1}^{\infty} \delta_i = \delta$. He considered certain finite coverings $H^i = \bigcup_k H_k^i$, calling them *Harnack coverings*, as follows. Cover G_1 by a finite number of closed intervals H_k^1, whose endpoints are not in G, such that $|H^1| < \mu_H(G_1) + \delta_1$. This step uses Osgood's assumption that G is nowhere dense. Cover $(G_2 \backslash H^1)$ by a finite number of closed intervals H_k^2, whose endpoints are again not in G, such that

$$|H^2| < \mu_H(G_2 \backslash H^1) + \delta_2$$

and hence:

$$|H^1| + |H^2| < \mu_H(G_2) + \delta_1 + \delta_2.$$

Continue in this fashion to obtain a sequence of coverings such that

$$|H^1| + |H^2| + \cdots + |H^i| < \mu_H(G_i) + \delta_1 + \delta_2 + \cdots + \delta_i$$
$$< \lim_{i \to \infty} \mu_H(G_i) + \delta.$$

After a finite number, i, of steps, all the points of G are enclosed in one of the H-intervals. For suppose the process were infinite. Then there would be at least one limit point of the endpoints of the intervals H_k^i. But there is a point of G within distance δ_i of an endpoint of H_k^i. These points form a sequence of points of G converging to the same point x' as the endpoints just mentioned. G contains its limit points, so x' must be a point of G and hence some

---
[32] See Hawkins, 1970, p. 115.

G_j. Since x' is in the interior of H_k^j for some k, a contradiction results. Osgood had again used a Bolzano-Weierstrass argument. Thus:

$$\mu_H(G) \leq |H^1| + |H^2| + \cdots + |H^i|$$
$$\leq \mu_H(G_i) + \delta_1 + \delta_2 + \cdots + \delta_i$$
$$< \lim_{i \to \infty} \mu_H(G_i) + \delta$$

Osgood's proof can be readily modified so that the nowhere dense restriction on the set G is not necessary, but Osgood either was unaware of this fact, or deemed it irrelevant to his main purpose.

As further preparation for his proof of the main result of Part II, Osgood refined Harnack's procedure for computing the content of a set G by first finding an interior content, μ^{int}, of the complement.[33] Perhaps Osgood noted that Harnack's procedure does not exhaust the complement if open intervals are removed whose endpoints do not lie in G. By requiring that the set G be closed, Osgood was able to improve Harnack's procedure by removing maximal open intervals at each step. Following Harnack, he then showed that if L is the length of an interval containing G, then $\mu_H(G) = L - \mu^{int}(G^{comp})$. Osgood's discussion is awkward and hard to follow, but the result is useful since Harnack content is not additive. (Recalling the earlier example, the Harnack content of the rationals in the unit interval is 1, as is the Harnack content of the irrationals, as is the Harnack content of the entire interval.)

With his convergence lemma and Fundamental Theorem in place, Osgood was now in a position to prove his main result. He was primarily concerned with the area of S_n above the line $y = A$ and did not concern himself about the area below $y = A$. This amounts to setting

$$S_n(x) = S_n^*(x) + S_n^{**}(x)$$

---

[33] See Osgood, 1897a, p. 177.

3. Term by Term Integration

where
$$S_n^*(x) = \begin{cases} S_n(x) - A, & S_n(x) > A \\ 0, & S_n(x) \leq A \end{cases}$$

and
$$S_n^{**}(x) = \begin{cases} A, & S_n(x) \geq A \\ S_n(x), & S_n(x) < A \end{cases}$$

Since both of these functions are continuous, they are integrable and
$$\int_{x_0}^{x_1} S_n(x)dx = \int_{x_0}^{x_1} S_n^*(x)dx + \int_{x_0}^{x_1} S_n^{**}(x)dx.$$

The size of the last term is controlled by the fact that it is less than or equal to $A(x_1 - x_0)$ so it is necessary only to show that the integral involving S_n^* converges to zero, and this is what Osgood did. The set G he selected for the application of his convergence lemma is the set of γ-points corresponding to A.

Osgood's Fundamental Theorem shows that G is closed and nowhere dense. Let
$$D_n = \{x \in [x_0, x_1] : S_j(x) \leq A \ \forall j \geq n\}$$
so the union of the D_n is the interval $[x_0, x_1]$. Then, the G_n of the convergence lemma are $G_n = D_n \cap G$.

Osgood wanted to show that given $\varepsilon > 0$, there exists an m such that
$$-\varepsilon < \int_{x_0}^{x_1} S_n(x)dx < \varepsilon$$
for all $n > m$. He proceeded by contradiction and considered the right hand inequality, for example. Assume that there is an infinite sequence $\{n_i\}$ for which, by choice of $A < \varepsilon(x_1 - x_0)^{-1}$, we have
$$\int_{x_0}^{x_1} S_{n_i}(x)dx \geq \varepsilon > A(x_1 - x_0), i = 1, 2, \ldots$$

Osgood showed that this is impossible by proving that the area under the curve $S_n(x)$ between A and the uniform bound B is less than ε_1, where $\varepsilon_1 < \varepsilon - A(x_1 - x_0)$, for all n sufficiently large. He called this area C_n.

Osgood took a Harnack covering H of G and proceeded as in the convergence lemma to take Harnack coverings H^n of G_n where $H^1 \subseteq H^2 \subseteq \cdots \subseteq H^n$. With the help of

$$\int_{H^n} S_n^* \leq (B - A)(|H^n| - \mu_H(G_n))$$

he showed that

$$C_n < (B - A)(\mu_H(G) - \mu_H(G_n) + \delta + \delta_1) \qquad (A)$$

where δ and δ_1 are positive constants such that

$$|H| < \mu_H(G) + \delta$$

and

$$|H^n| < \mu_H(G_n) + \delta_1$$

for all sufficiently large n. By his convergence lemma, there is a positive constant η such that $(\mu_H(G) - \mu_H(G_n)) < \eta$ for n large enough. Choosing the constants such that

$$\eta + \delta + \delta_1 < \frac{\varepsilon_1}{B - A},$$

he obtained the desired result:

$$C_n < (B - A)(\eta + \delta + \delta_1) < \varepsilon_1.$$

Otherwise stated,

$$\int_{x_0}^{x_1} S_n^*(x)dx$$

converges to zero as $n \to \infty$.

It is rather puzzling that Osgood chose to use Harnack coverings H^n of the sets G_n. It seems almost as if he felt he needed to reprove the convergence lemma. He did not need to use these coverings, since he could have shown directly that

$$\int_H S_n^* \leq (B-A)(|H| - \mu_H(G_n))$$

by exactly the same argument he used to establish (A). The result

$$C_n < (B-A)(\eta + \delta)$$

would have followed directly from the convergence lemma. Osgood's justification of (A) was based on approximating the integral by the use of lower Darboux sums, as he indicated by writing of "the limit approached by the sum of the inscribed rectangles."[34]

3.3.1 Conditions P

To some, it may seem strange that Osgood focused so much attention on nowhere dense sets. Throughout the paper Osgood referred to his *Conditions P*, which means closed and nowhere dense. He used Conditions P as a hypothesis for various results, but in the proof of a particular result, he sometimes used only one or the other of the two requirements. In his special case of Baire's theorem, Osgood used only the requirement that the component sets be nowhere dense. In his discussion on calculating the content of G by exhausting the complement, he used only the requirement that G be closed.[35] (In fact, the result is valid for general G. However, Osgood's process of removing maximal open intervals from the complement requires that G be closed.) In his convergence lemma, he used both Conditions P in his proof although the nowhere dense condition can be shown to be unnecessary.

Contemporaries of Osgood would not have been surprised by his focus on nowhere dense sets. It was originally believed, by

---
[34] See Osgood, 1897a, p. 187.
[35] See Osgood, 1897a, p. 177.

Hankel among others, that a nowhere dense set must have zero content. Then great attention was given to these sets following the discoveries by H.J.S. Smith in 1875 and Vito Volterra in 1881, of the existence of nowhere dense sets with positive content, as had been conjectured by Ulisse Dini. The set-theoretic properties of such sets pushed mathematicians in the direction of a theory of measure.[36] Osgood's use of Harnack's content may have contributed to his emphasis on closed sets. For his work in general, Osgood was surely aware of some of the differences in the properties of the Harnack content for open and closed sets. In particular, in modern terms, the Harnack content of a closed, bounded set is equal to its Lebesgue measure, while this is not necessarily true of an open set.

Osgood's exposition in this paper is not of the highest quality. In Part II, he essentially re-proved his convergence lemma in the body of the proof of his main result. In Part I, he gave more than once the proof of what is essentially his Fundamental Theorem and he also did not examine which of the Conditions P (closed, nowhere dense) he actually used for each result. Thus, in this paper, Osgood acted as a problem solver rather than a conceptualizer. Once he had his result, he did not spend time on simplifying the proof, reflecting on its essential character, or polishing his exposition. He seemed to lack skill in writing and exposition at this point in his career and would have benefited from the services of a good editor. It is also the case that the style and standards of mathematical writing of the time were quite different from the current ones, as a quick glance of the volumes of the American mathematical journals of 1897 will show. A mathematician in Osgood's position might also have found it difficult to write a paper to appear in an American journal that, on the one hand, would be accessible to a good number of his American colleagues and, on the other hand, would earn the respect of European mathematicians.

---

[36]See Hawkins, 1970, pp. 55–56. See Schoenflies, 1899, p. 89 for a discussion of nowhere dense sets. Smith's example is in Smith, 1875.

3.4 Reaction to Osgood's Paper and its Consequences

European mathematicians received Osgood's paper with interest at the time, and American and European mathematicians reflected on it in future years. The term by term integration theorem received comparisons with the work of Lebesgue and Borel. By the time Lebesgue wrote his thesis in 1902, he was aware of Osgood's work in this area and gave it prominent mention. In his introduction he wrote:

> Le calcul effectif d'une intégrale dépend essentiellement de la façon dont est donnée la fonction à intégrer. Dans le cas où la fonction est définie à l'aide de séries on pourra se servir de cette propriété, dont un cas particulier a été obtenu par Mr Osgood: Une série dont les termes ont des intégrales et dont les restes sont, en valeur absolue, inférieurs à un nombre fixe est intégrable terme à terme.[37]

Lebesgue again referred to Osgood's theorem before proving the above result in the body of his thesis, calling it the most interesting special case and commenting that Osgood obtained his proof "à l'aide de considérations toutes différentes."[38]

Émile Borel also became aware of Osgood's work. In his 1905 *Leçons sur les fonctions de variables réelles* (Lessons on functions of real variables, written up for publication by Maurice Fréchet and including sections by Paul Painlevé and Lebesgue), Borel introduced the subject of convergent series of real functions by describing Osgood's notion of examining the inversion of "double limits," citing

---

[37] See Lebesgue, 1902, p. 3. "The effective calculation of an integral essentially depends on the manner in which the function to integrate is given. In the case where the function is defined with series, one can use this property, of which a special case was obtained by Mr. Osgood: A series whose terms have integrals and whose remainders are, in absolute value, less than a fixed number can be integrated term by term."

[38] "...by a completely different method." See Lebesgue, 1902, p. 29.

88   3. Term by Term Integration

Osgood's earlier 1896 paper "A geometrical method for the treatment of uniform convergence and certain double limits."[39] Later in the same section, Borel remarked that the treatment of series integration is much simpler if it is approached from Lebesgue's point of view as opposed to Riemann's. He then stated the following bounded convergence theorem of Lebesgue:

> Si une série de fonctions bornées intégrables (L) est convergente de a à b et si l'on a $|r_n(x)| < M$ quels que soient l'entier n et l'abscisse x dans (a, b), la série des intégrales (L) des termes est aussi convergente et sa somme est l'intégrale (L) de $f(x)$.[40]

In a footnote, Borel, perhaps following Lebesgue, indicated that Osgood had already obtained this result for Riemann integration of continuous functions in a paper in the *American Journal of Mathematics*. (Osgood's paper is given as published in 1894 in Borel's book.)

Cesare Arzelà had stated a similar theorem for certain uniformly bounded sequences of Riemann integrable functions in 1885, giving a complicated combinatorial proof. Hawkins described Arzelà's condition on the mode of convergence as complicated and noted that his proof was extremely long and also complicated, as was Osgood's.[41] Borel also later discussed Arzelà's work in some detail.[42]

It is to Arzelà, not to Osgood, that Bourbaki gave credit for the result, which in Arzelà's case is more general in that he required the limit function to be Riemann integrable instead of continu-

---

[39] See citation in Borel, 1905, pp. 36.

[40] "If a series of bounded Lebesgue integrable functions is convergent between a and b and if we have $|s_n(x) - f(x)| < M$ for any whole number n and x in (a, b), the series of Lebesgue integrals of the terms is also convergent and the sum is the Lebesgue integral of $f(x)$." See also Borel, 1905, p. 49.

[41] See Hawkins, 1970. p. 117.

[42] See Borel, 1905.

3. Term by Term Integration

ous. When writing about Lebesgue's bounded convergence theorem, Bourbaki added a footnote with reference to Arzelà's paper:[43]

> Le cas particulier de ce théorème, où il s'agit d'une suite de fonctions intégrables au sens de Riemann dans un intervalle compact, uniformément bornées, et dont la limite est intégrable au sens de Riemann, avait été démontré par Arzelà.[44][45]

But Arzelà's work went largely unnoticed until Arthur Schoenflies brought it to light in a report which included a section on Osgood's paper. Schoenflies, who was on the faculty of the University of Königsberg, described Osgood's paper in his overview of the status of point set theory commissioned by the *Deutsche Mathematiker-Vereinigung*.[46] Some controversy apparently preceded this description. J.L. Walsh recalled the events as follows:

> Schoenflies wrote to Osgood, a much younger and less illustrious man, that he did not consider Osgood's results correct. The latter replied in the spirit that he was surprised at Schoenflies' remarkable procedure, to judge a paper without reading it. When Schoenflies' report appeared (1900), it devoted a number of pages to an exposition of Osgood's paper.[47]

Regardless of the details of this unconfirmed story, the important point is that Osgood's work became known to mathematicians in Europe by means of this widely read report. It is interesting to note that Schoenflies wrote of Osgood's convergence lemma without

---

[43] See Arzelà, 1885.
[44] "The special case of this theorem for a series of uniformly bounded Riemann integrable functions in a compact interval with a Riemann integrable limit was proven by Arzelà."
[45] See Bourbaki, 1960, p. 251.
[46] See Schoenflies, 1899, pp. 91, 227–233.
[47] See Walsh, 1989, p. 82.

indicating that Osgood's statement of the lemma was for nowhere dense sets.[48]

The reactions of Lebesgue, Borel and Schoenflies came within a few years after publication of Osgood's paper, but his results continued to receive some attention. In their classic *Functional Analysis*, Frigyes Riesz and Béla Sz.-Nagy, when introducing Lebesgue's theorem, felt it appropriate to mention the predecessors of it that belonged to the classical theory due to Arzelà and Osgood.[49] Bourbaki counted Osgood as a member of a group including Borel, Baire, Lebesgue and Young that accomplished important work in the classification of point sets. In particular, they gave Osgood credit for his special case of Baire's theorem, writing of the

> ...rôle fondamental joué en Analyse moderne par la notion d'ensemble maigre, et par le théorème sur l'intersection dénombrable d'ensembles ouverts partout denses dans un espace métrique complet, démontré d'abord (indépendamment) par Osgood pour la droite numérique et par Baire pour les espaces $\mathbb{R}^n$.[50]

In his *Semicentennial History of the American Mathematical Society*, R. C. Archibald stated that the result "was general and altogether new; and the ideas were akin to those which later led to Borel's definition of measure."[51] J.L. Walsh wrote that, for measurable functions, the result "became a model for Lebesgue in his new theory of integration."[52] Do these comments hit the mark? Was Osgood's work measure-theoretic to some extent? Did it serve as a model for Lebesgue's theory of integration? Archibald's comment

---

[48] See Schoenflies, 1899, p. 91.

[49] See Riesz and Sz.-Nagy, 1955, reprint 1990, p. 38.

[50] "...fundamental role in modern analysis played by the notion of nowhere dense set, and by the theorem about the countable intersection of dense open sets in a complete metric space, first proven (independently) by Osgood for the real line and by Baire for $\mathbb{R}^n$." See also Bourbaki, 1960, p. 177.

[51] See Archibald, 1938, p. 153.

[52] See Walsh, 1989, p. 82.

misses the mark somewhat. Although hints of a measure-theoretic approach are seen in Osgood's use of Harnack's content and in his convergence lemma, his basic method of proof depends on extraneous topological considerations (the use of nowhere dense sets). The critical measure theoretic breakthrough would come with Borel's use of infinite coverings, rather than the finite coverings used by Osgood. Walsh's assessment perhaps gives a somewhat too general impression. Lebesgue did, however, consider Osgood's result the most interesting special case of the theorem, and clearly believed it was significant. It appears that Osgood was Lebesgue's model for this particular important theorem in his theory of integration.

Norbert Wiener, a younger colleague of Osgood at Harvard and a man of no small ego himself, also offered his reaction to Osgood's work. Wiener gave some characteristically double-edged and grudging praise when, presumably referring primarily to this result, he wrote in his autobiography:

> He [Osgood] had done able work in analysis, against the resistance of those inhibitions that perpetually drive a certain type of New Englander from the original into the derivative and the conventional. Some of his ideas should have led him to the discovery of the Lebesgue integral, but he had not brought himself to the final step which might have led him to accept the striking consequences of his own conception. He must have had some rankling awareness of how he had missed the boat, for in his later years he would never allow any student of his to make use of the Lebesgue methods.[53]

Whether or not Wiener was accurate in his characterization of Osgood's stance toward the Lebesgue integral, this story has certainly become entrenched in mathematical lore. If it is true that Osgood's

---

[53] See Wiener, 1964, pp. 231–232.

students did not use the Lebesgue integral, it should also be noted that it was not altogether common for students anywhere to use the Lebesgue integral during Osgood's years at Harvard. It seems clear that the ideas that Wiener felt ought to have led Osgood to the discovery of the Lebesgue integral were the ideas contained in the paper under discussion in this chapter. Yet instead of having had "some rankling awareness of how he had missed the boat," Osgood may well have been content to have been in the mainstream of one of the most important mathematical dialogues of the time. He did not miss the boat—he just was not the only person on it.

Osgood's 1897 paper received favorable attention in Europe, by some of the best mathematicians of the day. His work became a part of a mainstream mathematical conversation on a topic of wide interest in analysis, involving among others Harnack, Schoenflies, Lebesgue and Borel. This represented a major difference between Osgood and his predecessors at Harvard. Benjamin Peirce's mathematical research was outside the mainstream and largely unknown in Europe until after his death. He was not a contributing member of the larger mathematical community that included Europe. Osgood's generation of American mathematicians established the American mathematical community's commitment to research. The result on term by term integration represented Osgood's first major contribution to that process. In terms of the history of ideas, the passage of time has shown that his main result was an important part of the development of the theory of integration, though superseded by Lebesgue's results. His Baire category results were of more enduring interest. Osgood's next major contribution to mainstream mathematical research tackled an even more high profile problem—the Riemann mapping theorem.

Chapter 4

Osgood's Proof of the Riemann Mapping Theorem—an "Outstanding Result"

Osgood's proof of the Riemann mapping theorem appeared in 1900 in the very first volume of the *Transactions of the AMS*, which was edited by E.H. Moore, E.W. Brown and T.S. Fiske. J.L. Walsh, in an account of Osgood's work, refers to this as his most outstanding single result.[1] The paper was entitled "On the Existence of the Green's function for the most general simply connected plane region."

---

[1] See Walsh, 1989, p. 82.

4. Osgood's Proof of the Riemann Mapping Theorem

4.1 Setting and Structure of Osgood's Paper

4.1.1 The Green's function—Why its existence establishes the Riemann mapping theorem

The Green's function for a simply connected region T is a function u satisfying the following conditions: u is harmonic[2] in T except at an arbitrary point O in the interior of T; is of the form $u = \log(1/r) + w$, where r is the distance to the point O and w is harmonic in T; and approaches zero at boundary points of T. Once the existence of such a function u has been established, a one-to-one analytic function f from T onto the interior of the unit circle can be obtained. Osgood assumed that his readers would supply the details of this procedure, which is outlined as follows. Let v be a harmonic conjugate of w and define $f(z)$ to be $e^{-\log|z|-w-iv}$ where w and v are functions of the complex variable z. It can then be shown, by standard techniques, that f is a one-to-one analytic function from T into the interior D of the unit circle.[3] Moreover, it can be shown that since $|f(z)|$ approaches one at the boundary points of T, the range of f is all of D. Thus the version of the Riemann mapping theorem that Osgood proved says that any simply connected region, with at least two boundary points, can be mapped conformally onto the interior of the unit circle.

4.1.2 The link with potential theory

The setting for study of the Green's function is potential theory, a subject that interested Osgood's teacher at Harvard, mathematical

---

[2] *Simply connected* means that any simple closed curve in the region can be shrunk continuously into a point, all the while remaining in the region. The simplest example: a disk is simply connected, but a washer is not. A harmonic function satisfies the Laplacian $\Delta u = 0$.

[3] See, for example, Osgood's own use of these standard techniques in a text he wrote in 1936: *Functions of a Complex Variable*, New York, Hafner, 1948, pp. 231–233.

4. Osgood's Proof of the Riemann Mapping Theorem

physicist B.O. Peirce, author of a textbook on the Newtonian potential function.[4] It was also one of the various interests of Felix Klein. Although it was Osgood's colleague Bôcher who attended Klein's lectures on potential theory at Göttingen, Klein's work in the field almost certainly stimulated Osgood's own.

The problem of finding a Green's function for a given region is a special case of what is known as the first boundary value problem of potential theory. Osgood likely learned of the Green's function as a student in the context of its applications to physics. In his textbook on functions of a complex variable[5] written late in life, he offered students this physical evidence for the existence of the Green's functions:

> Why is it reasonable to suppose that there will be any function which will fulfil the conditions which define the Green's Function? The physical evidence is highly convincing. It is as follows. Conceive a piece of tin foil cut out in the form of the region S. Let the plate be connected along with its whole edge to a thick piece of copper, forming a ring about it. If, now, the two poles of a battery are connected, one with the plate at the point O, the other with the copper, a flow of electricity will be set up, in which the electricity will flow through the plate from O to the edge. Now, the resistance of the copper is negligible, compared with that of the plate, and so the potential of the copper is sensibly constant; take it as 0.
>
> On the other hand, the potential near O is high, and rises to great height along a small circle about O. If O were truly a point source, it would become infinite.
>
> Thus the potential at all inner points of the plate but O satisfies Laplace's equation; it vanishes on the boundary; and it becomes infinite at O. Now a harmonic function

---
[4] See Peirce, 1888.
[5] See Osgood, 1948.

96   4. Osgood's Proof of the Riemann Mapping Theorem

which becomes infinite at a point is of the form:

$$c \log \tfrac{1}{r} + \omega(x, y),$$

where $\omega(x, y)$ is harmonic at the point. Hence the physical evidence for the existence of a Green's Function for the region is complete.[6]

Osgood was sometimes inspired by physical intuition, as in the example above, and he was always interested in physics, even to the point of writing a textbook on mechanics late in his career.[7] He was, however, a careful mathematician and did not make the mistake of assuming, based on physical evidence, the existence of the Green's function for an arbitrary simply connected region. Klein, Osgood's mentor at Göttingen, interpreted some of Riemann's work on Riemann surfaces in terms of physics and he was for a time mistakenly convinced that potential theory and current flows on surfaces were the inspirations behind Riemann's insights.[8] Although this was apparently not the case for Riemann, potential theory and current flow were of compelling interest to Osgood.[9]

4.1.3 Structure of the Green's function paper

Osgood's paper establishing the existence of the Green's function for the most general simply connected region is only four-and-a-half pages long, not counting a one-and-a half page note giving further details that Osgood published in the *Notes and Errata* section of the 1901 volume of the *Transactions*. It is loosely divided into three parts. The first is composed of introductory remarks. Here Osgood stated the existence theorem, gave a few sentences about the history

---
[6]See Osgood, 1948, p. 206.
[7]See Osgood, 1937.
[8]See Parshall and Rowe, 1994, pp. 178–179.
[9]For a succinct account of Klein's interest in Riemann's work, see Parshall and Rowe, 1994, chapter 4.

4. Osgood's Proof of the Riemann Mapping Theorem

of progress toward the most general proof of the Riemann mapping theorem, and developed an example to show just how "bad" a simply connected region can be. The next two parts of the paper give the proof itself, with one devoted to the case of a region of finite extent, and the other to the general case, including unbounded regions.

Osgood's structuring of the paper in this way is curious. His division of the proof into two cases, essentially bounded and unbounded, may mislead the reader into believing that the boundedness of the region is an essential factor in the proof of the theorem. He used the bounded case for purposes of preliminary exposition. To achieve his final result, however, he dropped the consideration of boundedness. Actually, for proving the Riemann mapping theorem boundedness is not an issue because there is a simple trick that reduces the unbounded case to the bounded case i.e., it is not hard to find a conformal mapping of an unbounded T onto a bounded region T_1.[10]

It is therefore unnecessary to divide the proof into two cases and it seems likely that Osgood would have known this. But Osgood offered no explanation of his own for why he structured the paper in this way. As the term by term integration paper demonstrated, Osgood was not a master of clear exposition in his early research papers. In this case, however, he may also have been reflecting the fact that Axel Harnack addressed only the bounded case in his 1887 work toward a proof of the Riemann mapping theorem, from which Osgood got a crucial idea.[11]

---

[10] See *Conformal Mapping* by Zeev Nehari for one description of the method for finding this conformal mapping. See also Nehari, 1975, pp. 175–176.

[11] Harnack's method is discussed in some detail later in this chapter.

4.2 Historical Context—It was "in the air"

Around 1900, complex analysis was considered one of the most important areas of mathematics. If a mathematician aimed to do something significant, he might have attempted an improvement of Picard's theorem or worked on one of the few other central issues in the theory of functions.[12] Osgood followed exactly this strategy—he gave his groundbreaking 1898 AMS Colloquium Lectures on the theory of functions, and his first topic was Picard's theorem.[13] Osgood considered this restricted form of Picard's theorem: "Any function $G(z)$ which is single valued and analytic for all finite values of z takes on in general for at least one value of z any arbitrarily assigned value c. There may be one value, a, which the function does not take on. But if there is a second such value, b, the function reduces to a constant."

Osgood traced the steps of Picard's proof of a restricted version of the theorem, noted that it was "intuitional in the use it makes of continuous deformations of continuous curves," and placed himself firmly in the camp of Weierstrassian rigor by stating that to "place the proof on the firmest foundation we possess it remains to arithmetize these steps," which he proceeded to do.[14]

The Riemann mapping theorem was another topic at the center of attention—it was "in the air." Karl Weierstrass brought attention to the theorem in 1870 when he pointed out a gap in Riemann's proof of it. Weierstrass's student H.A. Schwarz, Axel Harnack, Paul Paraf and others attempted to find a rigorous proof of the theorem, and succeeded for various special cases. As a topic at the center of attention, it too was a topic of interest in Osgood's AMS Colloquium Lectures. In these lectures, Osgood was already, two

---

[12]This assertion is based on the author's conversation with Saunders Mac Lane, February 1996.

[13]See "Selected Topics in the General Theory of Functions, Lecture 1," Bulletin of the AMS, vol. 5, no. 1, October 1898, page 59.

[14]See Osgood, 1898, p. 60.

4. Osgood's Proof of the Riemann Mapping Theorem

years before the publication of this Green's function paper, mulling over some results by Harnack and Poincaré with a view to reaching a general proof of the Riemann mapping theorem. Lectures II and III in the series[15] were devoted to an 1883 paper of Poincaré, with mention of related work by Harnack. Osgood demonstrated a connection between Poincaré's results and the Riemann mapping theorem by noting that if the theorem were true, the existence of a function with certain properties Poincaré described would be assured. He went on to state: "The determination of the correctness or incorrectness of this [Riemann mapping] theorem would be a useful contribution to analysis."[16] Perhaps it was one he hoped to make.

4.2.1 Riemann's statement and "proof"

The statement of the Riemann mapping theorem can be traced to Bernhard Riemann's 26 December 1851 inaugural dissertation "Grundlagen für eine allgemeine Theorie der Functionen einer veränderlichen complexen Grösse" (Foundations for a general theory of functions of a complex variable). This famous dissertation stimulated much basic research in the theory of analytic functions, topology, algebraic geometry and differential geometry. Writing in 1951 for a conference commemorating the 100th year of Riemann's dissertation, Lars Ahlfors declared that "the most astonishing feature in Riemann's paper is the breathtaking generality with which he attacks the problem of conformal mapping." Ahlfors gave the Riemann mapping theorem as an example but stated that in Riemann's dissertation it "is ultimately formulated in terms which would defy any attempt of proof."[17] Later generations of mathematicians then formulated the theorem for various types of regions of the complex plane and sought proofs.

---

[15] See Osgood, 1898, pp. 69–74.
[16] See Osgood, 1898, p. 74.
[17] See Ahlfors, 1953, pp. 3–4.

4. Osgood's Proof of the Riemann Mapping Theorem

In his dissertation, Riemann formulated his theorem as follows:

> Zwei gegebene einfach zusammenhängende ebene Flächen können stets so auf einander bezogen werden, dass jedem Punkte der einen Ein mit ihm stetig fortrückender Punkt der andern entspricht und ihre entsprechenden kleinsten Theile ähnlich sind; und zwar kann zu Einem innern Punkte und zu Einem Begrenzungspunkte der entsprechende beliebig gegeben werden; dadurch aber ist für alle Punkte die Beziehung bestimmt.[18]

Riemann himself justified his mapping theorem using the existence of the Green's function. Thus Osgood followed Riemann in employing this method from potential theory. But Riemann assumed the so-called Dirichlet principle which left a gap in his proof subsequently pointed out by Karl Weierstrass.

4.2.2 The Dirichlet principle/problem and the critique made by Weierstrass

P.G Lejeune Dirichlet formulated his principle in lectures on potential theory given at Göttingen in 1856–57 as follows:

> For an arbitrary bounded domain there is always one and only one function u of x, y, z which, together with its first-order derivatives, is continuous, satisfies the equation $[\Delta u = 0]$ within this entire domain and reduces to a given value at every point of the [boundary] surface.[19]

---

[18]Riemann's formulation is also quoted in Hille, 1973, p. 320. Loosely translated: "Two given simply connected plane regions can (always) be related to one another in such a way that to each point of the first corresponds a point of the second in a continuous manner and their corresponding smallest parts are similar; and in fact to an interior point of the one and to a boundary point of the one the corresponding (points of the other) can be assigned arbitrarily; but by this the relation is determined." See Riemann, 1876, p. 40.

[19]Quoted from the translation in Bottazzini, 1986, page 299. Bottazzini's book contains an appendix "On the History of Dirichlet's Principle," pages 295–303, which describes the history

4. Osgood's Proof of the Riemann Mapping Theorem

The Dirichlet principle deals with the *existence* of such a function; the Dirichlet "problem" then has to do with determining what that function is, i.e., with finding the function u that is harmonic in the interior of the domain and continuously takes on given boundary values. The Dirichlet problem in two dimensions can be formulated as one of finding the function u that minimizes the integral equation

$$U = \iint_D \left\{ \left(\frac{\partial u}{\partial x}\right)^2 + \left(\frac{\partial u}{\partial y}\right)^2 \right\} dxdy.$$

This was a characteristic problem from the calculus of variations, a field in which Osgood was expert. Riemann, a student of Dirichlet, gave the Dirichlet principle its name. Following his teacher, Riemann assumed that since $U \geq 0$ there would exist a function u that minimized U and possessed the required properties.[20]

In 1870, Weierstrass raised an objection, on the grounds not only of rigor but also of correctness.[21] In his critique of what he termed the "so-called" Dirichlet Principle,[22] he pointed out that in the calculus of variations it was known that the problem of minimizing an integral of the above character need not have a solution, i.e., that the greatest lower bound need not be achieved by a function with the same properties as the set of functions u_n used to obtain it.[23] In other words, it was not proven mathematically that the minimal function u solved the Dirichlet problem, even though the members of any sequence of functions u_n converging to u did meet the necessary criteria (i.e., they were continuous with continuous first order derivatives, took on the given boundary values and were harmonic).

---

in more detail. For the full story, see Monna, A.F.: *Dirichlet's Principle: A Mathematical Comedy of Errors and Its Influences on the Development of Analysis*. Utrecht, Oosthoeck, Scheltema, and Holkema, 1975. See also Dirichlet, 1887, p. 127.

[20] See Monna, 1975, p. 33.
[21] See Weierstrass, 1870 in *Mathematische Werke*, v. 2, pp. 49–54.
[22] In fact the title of the paper is "Über das Sogenannte Dirichlet'sche Princip."
[23] See also Bottazzini, 1986, p. 300.

In fact, Weierstrass gave an example where the limit function u was discontinuous. In his example[24]

$$J = \int_{-1}^{1} x^2 \left[\frac{dy}{dx}\right]^2 dx$$

where

$$y = \begin{cases} a & x = -1 \\ \dfrac{a+b}{2} + \dfrac{b-a}{2} \dfrac{\arctan(x/\varepsilon)}{\arctan(1/\varepsilon)} & -1 < x < 1 \\ b & x = 1. \end{cases}$$

The greatest lower bound for y is 0 and as $\varepsilon \to 0$

$$y = \begin{cases} b & x > 0 \\ \dfrac{a+b}{2} & x = 0 \\ a & x < 0. \end{cases}$$

Some physicists were not impressed with Weierstrass's critique. The physicist Hermann Helmholtz once told Felix Klein: "Für uns Physiker bleibt das Dirichletsche Prinzip ein Beweis."[25] This reflected the idea that physicists were less concerned with mathematical technicalities and were satisfied that the principle did indeed correspond to physical reality. Osgood was careful in his proof not to make the mistake of Dirichlet and then Riemann. He showed himself to be among the followers of Weierstrass by proving the existence of the limit function with the required Green's function properties.

---

[24] See Bottazzini, 1986, pp. 300–301, or Weierstrass, 1870, p. 53.
[25] See Klein, 1967, v. 1, p. 264. "For us physicists, the Dirichlet Principle remains a proof."

4. Osgood's Proof of the Riemann Mapping Theorem

4.2.3 Steps toward providing a rigorous proof of the Riemann mapping theorem

Hermann Amandus Schwarz, a student of Weierstrass, worked on correcting the error that his professor had pointed out in Riemann's proof by finding cases of the mapping theorem which were true. Osgood gave Schwarz credit for proving the theorem for regions bounded by a finite number of pieces of analytic curves. He also remarked on the work of Amédée Paraf and Paul Painlevé who proved the theorem in the case of regions bounded by curves with continuously turning tangents.[26]

Axel Harnack attempted, not entirely successfully, to take a next step in 1887 by proving the existence of the Green's function for an even more general region than those considered by Schwarz, Paraf and Painlevé, one bounded by an arbitrary Jordan (simple closed) curve.

4.3 Relationship of Osgood's Proof to the Work of Poincaré and Harnack—Two Primary Influences

Osgood had continued to follow Harnack's work—recall that he based part of his term by term integration paper on Harnack's use of content—and again used some of Harnack's methods, as well as methods due to Poincaré, in this Riemann mapping theorem paper. In his 1901 note in the *Transactions*, Osgood clarified to some extent the relationship of his work to Harnack's in the latter's 1887 *Die Grundlagen der Theorie des logarithmischen Potentiales und der eindeutigen Potentialfunktion in der Ebene*[27] and Poincaré's 1883 paper "Sur un Théorème de la Théorie générale des fonctions."[28]

---

[26] See Osgood, 1900, p. 310.
[27] See Harnack, 1887.
[28] See Poincaré, 1883.

He explained that his method and Harnack's (both substantially following Poincaré) are the same up to construction of a majorizing function U used to establish convergence of a sequence of Green's functions.

4.3.1 Poincaré's methods

The work that attracted the attention of Harnack and Osgood was contained in a May 1883 paper "Sur un Théorème de la Théorie générale des fonctions."[29] Poincaré proposed: "Soit y une fonction analytique quelconque de x, non uniforme. On peut toujours trouver une variable z telle que x et y soient fonctions uniformes de z."[30] Once unaware of the Dirichlet principle, he now knew of it and stated:

> Pour démontrer ce résultat, je me servirai du beau théorème de M. Schwarz (*Monatsberichte*, octobre 1870), connu sous le nom de *Principe de Dirichlet*.[31]

To prove his theorem, Poincaré started out with a given Riemann surface,[32] then constructed an infinite sequence of expanding closed curves $\{C_n\}$ such that each point of the Riemann surface was interior to one of the C_n. These curves were constructed in such a way that he could then assert that each region bounded by a C_n had a Green's function $u_n(z)$ (with common logarithmic singularity).[33] He then defined a function[34] ϕ which he used to construct another function $t(z)$ that majorized each $u_n(z)$, i.e., for fixed z,

---

[29] See Poincaré, 1883 On a Theorem from the General Theory of Functions.

[30] "Let y be any analytic function of x, non-uniform. One can always find a variable z such that x and y are uniform functions of z." See Poincaré, 1883, p. 112.

[31] "To prove this result, I will use the beautiful theorem of M. Schwarz (*Monatsberichte*, October 1870) known by the name *Dirichlet Principle*. See Poincaré, 1883, p. 113.

[32] For the proofs of Harnack and Osgood it was irrelevant that Poincaré used a Riemann surface.

[33] Poincaré described the "Green's function" but did not use that term himself.

[34] See Poincaré, 1883, p. 116.

4. Osgood's Proof of the Riemann Mapping Theorem

$$u_1(z) \leq u_2(z) \leq \cdots \leq u_n(z) \cdots \leq t(z)$$

He used $t(z)$ to show the existence of $u(z) = \lim_{n \to \infty} u_n(z)$ away from the logarithmic singularity of the Green's functions. Poincaré then demonstrated the continuity of u, the uniform convergence of the u_n to u, and concluded that this limit function is harmonic (again, in small neighborhoods away from the point where the Green's functions are logarithmically infinite).[35] Poincaré did not show that u approached zero on the boundary. The process of constructing a majorizing function was the key to the methods later used by Harnack and Osgood in their proofs of the Riemann mapping theorem.

Poincaré used, as noted above, a certain function ϕ to define his majorizing function t. He began this process somewhat cryptically when he wrote:

> Soient k le module d'une fonction elliptique, K et K' ses periodes"[36]

He defined ϕ using these periods K and K', but offered few additional details about ϕ. Osgood alluded to this unclarity about ϕ in the concluding paragraph of his Green's function paper by writing: "The theorem of this [Osgood's own] paper justifies the conclusion that the function $\phi(t)$ employed by Poincaré may be taken as an automorphic function."[37] Poincaré's ϕ is apparently what is customarily termed the modular function.[38] As mentioned above, he used this modular function to construct his majorizing function for a sequence u_n of Green's functions, although he did not use the term

---

[35] See Poincaré, 1883, pp. 117–120.
[36] "Let k be the modulus of an elliptic function, K and K' its periods." See Poincaré, 1883, p. 115.
[37] See Osgood, 1900, p. 314.
[38] The modular function is described in Ahlfors, 1979, pp. 277–181 where it is indeed shown that it is an automorphic function. Ahlfors wrote that "a function which is automorphic with respect to a subgroup of a modular group is called a *modular function* (or an *elliptic modular function*)."

"Green's function,"[39] and obtained a limit function u. He showed that u is harmonic and has logarithmic singularity at O.

From June 1881 through September 1882, Felix Klein and Henri Poincaré corresponded on exciting new developments from their work on what came to be called automorphic functions.[40] In 1881, Poincaré had the idea of what he called Fuchsian functions having been inspired by work in linear differential equations by Lazarus Fuchs. Felix Klein took exception to this name, believing that Schwarz had precedence. Poincaré did not disagree, but he had not known of a prior claim until after the name Fuchsian had already been bestowed. He was unwilling to insult Fuchs with a renaming and instead named another type of function Kleinian. Eventually the dispute cooled.

Among the automorphic functions, the Fuchsian functions are invariant under discontinuous (i.e. discrete) groups of linear transformations that leave a circle fixed. These functions were useful in solving linear differential equations with algebraic coefficients.[41] Poincaré wrote a series of Fuchsian function papers between 1881 and 1883. Thus this 1883 paper was part of Poincaré's exploration of automorphic functions and their applications.

4.3.2 Harnack's method

Poincaré's paper inspired Axel Harnack's proof of a restricted case of the Riemann mapping theorem.[42] Harnack began his proof by reminding the reader that the Green's function for an arbitrary

---

[39] See Poincaré, 1883, p. 116.
[40] The entire correspondence is included in Poincaré's collected works—Poincaré, 1956, v. XI, pp. 26–65).
[41] See Gray, 1986, p. 277, and Gray, 1986, Chapter VI, pp. 246–316 for a full account of the works of Klein and Poincaré related to automorphic functions and a summary of their correspondence.
[42] See Harnack, 1887.

4. Osgood's Proof of the Riemann Mapping Theorem 107

polygon was known to exist.[43] He then offered a careful explanation of the term *einfach zusammenhängende* (simply connected) and specified that he would consider regions bounded by a Jordan curve, C.[44] Next he constructed an expanding sequence of polygonal regions whose union is the region T bounded by C. Like Poincaré, Harnack used the Green's functions for these polygons and a function, $\log(R/r)$ specifically, that majorized them in order to prove the existence of a harmonic limit function, his candidate for the Green's function. Then he needed to find a method to show that his limit function u approached zero on the boundary. Harnack's important idea consisted of constructing another customized majorizing function for each boundary point, a. To do this, he let Q be a closed polygonal region whose only points of intersection with T were boundary points, including especially the boundary point a. He used the Green's function U for this polygon as his customized majorizing function for the boundary point a. Since $0 \leqslant u_n \leqslant U$, it followed that $0 \leqslant u \leqslant U$. Then the fact that $U(z) \to 0$ as $z \to a$ implied also that $u(z) \to 0$ as $z \to a$. He concluded:

> Sonach ist die Existenz der Greenschen Funktion für eine geschlossene einfach zusammenhängende beliebig berandete Fläche gezeigt.[45]

In summary, Poincaré's work showed Harnack a method for finding a majorizing Green's function that ensured the existence of a harmonic limit function u with logarithmic singularity at O. Harnack then took the method a step further, constructing a customized majorizing function for each boundary point that would ensure that the limit function u was itself a Green's function. Harnack thus showed that the problem of constructing a Green's function for a domain T would be solved if one could find, for each boundary point

---

[43] See Harnack, 1887, p. 116.

[44] Jordan curves are defined in the first paragraph of Chapter 5, which describes Osgood's discovery of a Jordan curve of positive area.

[45] "Thus the existence of the Green's function for an arbitrarily bounded closed simply connected region is shown. See Harnack, 1887, p. 121.

a, a majorizing function U which is harmonic except for a logarithmic singularity, positive in the domain, and such that $\lim_{z \to a} U(z) = 0$.

In his 1901 note, however, Osgood clarified how his proof improved on Harnack's. For some boundary points, a majorizing function U could not be constructed using Harnack's method. To quote Osgood:

> His [Harnack's] analysis suffices to show that the boundary function g (or u) will take on the required boundary value in the point A, but not that this will be the case for a point of the boundary of F that cannot be reached by a polygon Q. Thus an ordinary beak-shaped cusp (Schnabelspitze) could not be treated by Harnack's method. It appears, then, that Harnack did not solve the problem he proposed even for regions F bounded by a finite number of pieces of analytic curves, to say nothing of regions, some of the points of whose boundaries cannot be approached along a continuous curve lying wholly within F. In my solution, I have employed the same method of the majorante (the function U) adopted by Harnack, but have so chosen U that my proof covers *all* cases; and I have pointed out that there are here included cases which, I believe, have never been thought of before.[46]

In effect, Harnack assumed that for each point a on the boundary of T there is a polygon C such that T is contained in the interior of C and a is on the boundary of C. Osgood realized that the polygon C does not always exist and concluded that a more complicated majorizing function was needed.

Another result contained in Harnack's 1887 book was of interest to Osgood in writing his Riemann mapping theorem paper. Har-

---

[46] See Osgood, 1901, p. 485.

4. Osgood's Proof of the Riemann Mapping Theorem

nack improved—in a theorem now known as Harnack's theorem—on the technique Poincaré used in his paper to determine convergence of harmonic functions to a harmonic limit function.[47] Osgood paraphrased the theorem this way:

> ...if a function u_n is harmonic throughout a region T for all values of $n \geq m$ and if u_n increases at all points of the region when n increases; if furthermore at one single interior point of T, u_n approaches a limit when n becomes infinite, then u_n converges at all interior points of T and the limit u is harmonic throughout the interior of T.[48]

In terms of his sequence of Green's functions, this showed that in order to prove that $\lim_{n \to \infty} u_n(z)$ exists for all z in a neighborhood away from the logarithmic singularity, it suffices to show that this limit exists for *one* point z in that neighborhood.

Harnack knew of Poincaré's 1883 paper and credited the Frenchman in a footnote for his method of proof of the Riemann mapping theorem.[49] Harnack, however, did not recognize that his majorizing functions could not be constructed for some regions bounded by Jordan curves. Osgood recognized that the boundary of a region T, either one bounded by a Jordan curve or a more general region, could be so badly behaved that a more complicated method of arriving at a majorizing function was needed. He took the kernel of Poincaré's method of constructing a majorizing function using the automorphic function ϕ, and attacked the problem of the Riemann mapping theorem again.

---
[47] See Harnack, 1887, p. 67.
[48] See also Harnack, 1887, pp. 65–68, and Osgood, 1900, p. 312.
[49] See Harnack, 1887, p. 121.

4.4 Osgood's Proof

Clearly Osgood did not offer an altogether new proof of the Riemann mapping theorem from first principles; he worked with methods drawn from the work of Riemann, Poincaré, Schwarz and Harnack. Moreover proofs of various special cases of the theorem already existed well before 1900. With his customary care in crediting his sources, Osgood wrote some of the history of the theorem into his introductory remarks. As previously mentioned, Osgood noted that Schwarz had proved the existence of the Green's function for regions bounded by a finite number of pieces of analytic curves and that Paraf and Painlevé had done the same for regions bounded by curves with continuously turning tangents. Osgood also built on the work of his predecessors by using the already known existence of the Green's function for polygonal regions in his proof.

4.4.1 Osgood's main insight—Badly behaved boundaries

Osgood aimed to establish the existence of the Green's function for the most general simply connected plane region T. His insight lay principally in realizing just how badly the boundary of such a region could behave. As an example, his paper described a comb-like region constructed using a set of points on the x-axis that is perfect, nowhere dense and of positive content.[50] He cut out the perpendicular lines of unit length erected on the points of this set. Then he let T be the remaining portion of the upper half plane. This region T is similar to the rational comb that students learn of in introductory topology. (Osgood describes another such badly behaved simply connected region in his 1903 paper constructing a Jordan curve of positive area.) T is simply connected yet there are inaccessible points on the boundary of T, those which cannot be reached from the interior of T by means of a continuous curve. For

---

[50] See Osgood, 1900, p. 311.

4. Osgood's Proof of the Riemann Mapping Theorem

this reason Osgood had to provide a clearer statement of what it meant to say that the Green's function approaches 0 on the boundary of T. He thus specified that if a is a point on the boundary of T, if $p_1, p_2, p_3, \ldots$ lie in T and have a as their only limit point, and if u is the candidate Green's function, then $\lim_{n \to \infty} u(p_n) = 0$. (Osgood, 1900, p. 311) Osgood referred to this statement as condition (3) of the Green's function.

4.4.2 A summary of the proof

Osgood divided the proof into two parts, the bounded case and the unbounded case. As previously described, this division was unnecessary. In the bounded case, he constructed a nested sequence of polygonal regions (known to have Green's functions) in the interior of the region T. (This was similar to Harnack's use of polygonal regions in the interior of the region.) He then enclosed the region T in a circle with Green's function U and used Harnack's theorem to show that the u_n converged to some limit function u. It remained, however to show that condition (3) was fulfilled by this u, i.e., that $\lim_{n \to \infty} u(p_n) = 0$. As discussed earlier, Osgood did not complete this proof for the bounded case but proceeded directly to the general case. He employed the same interior polygons and their Green's functions as in the bounded case. As will be described below, he constructed, for a given boundary point a, a more complicated majorizing function U that established convergence to the required limit.

4.4.3 When T is a bounded region

Osgood's proof of the existence theorem began with the case in which T is a bounded region of the extended complex plane (i.e., including the point at infinity). He divided the plane into squares of side length 2^{-n}; for n sufficiently large, some of these squares lie in T. He then let C_n be the largest (simply connected) polygon that can be formed of such squares at stage n, and designated as u_n the Green's function for the region C_n with logarithmic singularity at the arbitrarily chosen interior point O. Basic properties of harmonic functions imply that as n increases, $u_n(z_0)$ increases for a given z_0 in T and remains positive since the minimum, zero, must be on the boundary.[51]

Osgood next let C be a circle of radius R and center O such that C contains the entire bounded region T; $U = \log(R/r)$ is the Green's function for C and $u = \lim_{n \to \infty} u_n$. This limit exists since the sequence $\{u_n(z_0)\}$ is bounded by $\log(R/r)$ and is increasing. Defined in this manner, u exhibits the required logarithmic behavior at O for a Green's function. Osgood then used Harnack's theorem to show that u is harmonic in T, except at O.[53] It remained to show that u satisfied condition (3).

At this point, however, Osgood abandoned the proof of the bounded case without completing it. He offered only this explanation: "This mode of proof is typical for the general case, but it is necessary to use a more complicated function to establish the convergence. Such a function we will now introduce, choosing it at once in such a way that it can be used to show that the limiting function u satisfies condition (3) too."[54] He proceeded directly to the general case and constructed a different majorizing function U (in his 1901 note, he called it a majorante) that established not only the existence of the harmonic function $u = \lim_{n \to \infty} u_n$ but also the convergence to zero of u at the boundary.[55]

---

[51] The maximum principle for harmonic functions states: "If the function $v(x, y)$ is harmonic in a domain D, it cannot attain its absolute maximum or minimum at an interior point of D unless $v(x, y)$ reduces to a constant." The following is a corollary: "If $v(x, y)$ is harmonic in a domain D and continuous in the closure of D, then both the maximum and minimum values of v in the closure of D are attained on the boundary."[52]

[53] See Osgood, 1900, p. 312.
[54] See Osgood, 1900, p. 312.
[55] See Osgood, 1900, pp. 312–313.

4. Osgood's Proof of the Riemann Mapping Theorem

Osgood's method and Harnack's coincided in essence, although not entirely in substance, up to this point in the paper, and Harnack considered only the bounded case. But Harnack's majorizing function U was inadequate for the task. So Osgood abandoned both Harnack's boundedness condition and his majorizing function to embark on the proof of the general case, giving the reader only an unclear and incomplete explanation of what he was doing.

4.4.4 When T is any region, bounded or unbounded—The heart of the proof

Osgood used a method of constructing a majorizing harmonic function, inspired by Poincaré (via Harnack), although Poincaré used his method to tackle the somewhat different but related type of uniformization problem described earlier. Harnack was also inspired by Poincaré, but his construction of a customized majorizing function U did not work in any case where for a boundary point a of T there did not exist a closed polygonal region C with T sharing only boundary points of C, and specifically including a on the boundary of C. Osgood therefore had to construct a majorizing U that overcame this difficulty, i.e., he had to construct, for a given point a on the boundary of T, a function U harmonic in T except at O with the required logarithmic behavior at O, positive in T and itself meeting the same condition (3) required of his Green's function u, i.e. if $p_1, p_2, p_3, \ldots$ in T converge to a, then $\lim_{n \leftarrow \infty} U(p_n) = 0$. Osgood used a variation of Poincaré's method to construct U, in the process filling in details that Poincaré omitted or did not describe quite clearly.

114   4. Osgood's Proof of the Riemann Mapping Theorem

4.4.5   Details of Osgood's proof, and details he omitted

Osgood chose three points a, b and c on the boundary of the region T, which lies in the extended z-plane, i.e., the plane with the point at infinity.[56] He assumed without comment that the point at infinity was not in T. In fact, if it were in T a preparatory transformation could move it to the boundary. Osgood then cut along the circular arc abc from a to c and remarked:

> Then the plane, thus bounded, can be mapped conformally on a quadrilateral of a second plane (the w-plane), this quadrilateral being formed by a triangle whose sides are arcs of circles tangent to each other at the vertices and an adjacent triangle formed by reflecting the first triangle in one of its sides.[57]

Neither in his Colloquium Lectures nor in his paper did Osgood give enough details to make it clear how this mapping should be constructed. One way to fill in the details follows.

Let Δ be the complement of the arc abc. By first mapping the arc abc onto a suitable portion of the unit circle via a composition of translation, scaling and rotation, and then applying a linear fractional transformation, a function f_1 can be obtained (analytic except for a pole) that maps arc abc onto the non-negative part of the x-axis with $f_1(a) = 0$, $f_1(c) = \infty$ and, after multiplying by the appropriate positive constant, $f_1(b) = 1$. The next step involves application of the branch of the square root function which maps the complement of the non-negative x-axis (which is then the image of Δ) onto the upper half plane, followed by the mapping $z \to -\frac{z-1}{z+1}$,

---
[56] It may make reading Osgood's paper easier to think of this extended z-plane as a Riemann sphere. Then his theorem shows that a simply connected open set on the Riemann sphere can be conformally mapped onto the unit disk. Thinking in this way, the north pole (point at infinity) is a boundary point of T if and only if T is unbounded.

[57] See Osgood, 1900, pp. 312–313.

4. Osgood's Proof of the Riemann Mapping Theorem

thus obtaining a function f_2 which maps the complement of the non-negative x-axis onto the open unit disk. The composition of f_1 and f_2 provides a one-to-one analytic mapping h from Δ onto the open unit disk. Moreover h extends to the boundary abc of Δ in such a manner that: i) h is continuous at a and at c with $h(a) = 1$, $h(c) = -1$; and ii) depending on whether z approaches b from one side of the arc abc or the other, $h(b)$ approaches either i or $-i$.

Then it is necessary to map conformally this open unit disk (now the image of Δ under h) onto the "quadrilateral" whose sides are arcs of circles orthogonal to the unit circle. Consider first the inverse of this mapping. By Schwarz's restricted version of the Riemann mapping theorem for regions bounded by analytic arcs (which was known to Osgood), there is a conformal mapping θ from the interior of the desired quadrilateral onto the open unit disk. Moreover, the arcs of the quadrilateral are mapped continuously onto portions of the unit circle. It was also known that there is a three parameter group of linear fractional transformations that take the unit disk onto itself. There are, therefore, only three degrees of freedom in deciding the mapping of the vertices. So we can specify that $\theta(i) = i$, $\theta(-i) = -i$ and, but then the value $\theta(-1)$ is completely determined. Therefore, the main difficulty in constructing the desired conformal mapping is to show that $\theta(-1) = -1$. To accomplish this, let $\theta_1(z) = \overline{\theta(\overline{z})}$. (It can be shown that this is an analytic function.) Then, $\theta_1(1) = 1$, $\theta_1(i) = i$ and $\theta_1(-i) = -i$. Note then that θ composed with θ_1 is the identity map since it maps the unit circle onto itself with three points fixed. This means that $\theta(z) = \theta_1(z)$, in particular $\theta(-1) = \theta_1(-1) = \overline{\theta_1(-1)}$, which is a complex number equal to its conjugate on the unit circle that is not 1, i.e., $\theta(-1) = -1$. To complete this step, let $f_3(z) = \theta^{-1}(z)$—this is a conformal mapping from the open unit disk onto the quadrilateral.

Now finally, the composition $f(z) = f_3(f_2(f_1(z)))$ maps Δ (Osgood's "region thus bounded") conformally onto the quadrilateral and Osgood's missing details are filled in. This is an example of a typical feature of this paper—simple comments by Osgood hide a world of details, not all of them completely elementary.

Osgood next filled up the unit disk by repeated reflections of the quadrilaterals. He again gave no details in the paper, writing only that it was necessary to "construct now by successive reflections the complete Riemann's surfaces that belong to the analytic function and its inverse, defined by the conformal transformation just considered."[58] Some of the missing detail may be supplied in this way. Define a function g for every point in the open unit disk. To begin, g is defined to be f^{-1} on the interior of the first quadrilateral. Next, g is extended by Schwarz reflection across the four boundaries of the first quadrilateral. To complete the definition of g, this process is then continued indefinitely.[59] This function g is the automorphic function used by Osgood. Although g is a many-to-one function, independence of path (monodromy) can be used to define a branch of g^{-1} on T (because T is simply connected). Monodromy ensures that the values of g^{-1} obtained in this process do not depend on the path taken.[60] This is not an entirely elementary detail.

At this point Osgood had the automorphic function needed to construct his majorizing function U. Let U' be the Green's function of the unit circle. Then the required majorizing harmonic function U is defined as $U(z) = U'(g^{-1}(z))$. Osgood wrote: "U is then a function single-valued and harmonic on the z-surface except at the one point O, where it becomes logarith-

---

[58] See Osgood, 1900, p. 313.

[59] This can be viewed as a Riemann surface. Osgood mentioned this but he is not very clear, and in fact the modern language for discussing Riemann surfaces had not yet developed. A leaf of the Riemann surface is determined by picking a point O in T and taking a branch of the function g^{-1}.

[60] A form of the Schwarz reflection principle is necessary to ensure analyticity of g across the boundaries after reflections of the quadrilateral. Schwarz reflection is a form of analytic continuation. See a discussion of Schwarz reflection in, for example, Lang, 1985, pp. 324–327. The monodromy theorem is stated in Lang, 1985, p. 333.

4. Osgood's Proof of the Riemann Mapping Theorem

mically infinite; and moreover" (here Osgood put the key idea in italics):

> U approaches the value 0 when the point (x, y) approaches any one of the points a, b, c along any path whatever, or, more generally, when (x, y) passes through the points of a set $p_1, p_2, \ldots$, lying at pleasure on the surface, but having as the only point about which they cluster one of the points a, b, c.[61]

Osgood's majorizing function U not only ensured the existence of the limit function u of the u_n, but also ensured that it satisfied Green's function condition (3) for the most general simply connected region of the plane. He explained it in this way:

> Finally, it can be shown by means of the function U that u approaches the boundary value 0 when the point (x, y) approaches a, or more generally, when a set of points $p_1, p_2, \ldots$ is so taken in T that $\lim p_n = a$, when $n = \infty$. For the corresponding values of U: $U_{p_1}, U_{p_2}, \ldots$, approach 0 as their limit, and $0 < u \leq U$.[62]

Osgood thus proved the Riemann mapping theorem for the most general simply connected region, in one fell swoop proving it for regions bounded by a Jordan curve and for more general open sets such as the region he described with the comb-like boundary.

---

[61] See Osgood, 1900, p. 313.
[62] See Osgood, 1900, p. 313.

118   4. Osgood's Proof of the Riemann Mapping Theorem

4.5   Reactions and Consequences

On the occasion of the AMS semicentennial, G.D. Birkhoff wrote an overview of the mathematical work done in the United States during those 50 years (1888–1938). In that article, he devoted a paragraph in the section on analysis to Osgood's proof of the Riemann mapping theorem, noting that it was Osgood's best known original work and calling it "extremely elegant and general." Birkhoff also brought to the reader's attention a 1903 result,[63] by Osgood in collaboration with his student E.H. Taylor that if a boundary point a is accessible from within T by a continuous path, then, under the conformal mapping f, if z_n is a sequence of points such that $z_n \to a$ as $n \to \infty$, then $\lim_{n\to\infty} f(z)$ exists, and might be called $f(a)$. Birkhoff paraphrased the theorem to say that a neighborhood of a boundary point accessible by a continuous path within the region T is mapped on the neighborhood of one point of the bounding circle.[64] Osgood announced the result in 1903, and Taylor read "a paper" on the subject at the April 1910 meeting of the AMS which was announced in the *Bulletin of the AMS* (v. 16, p. 452). The joint paper, however, that included theorems and proofs was not published until 1913.[65]

What was the significance of this further result? The Riemann mapping theorem maps the interior points of the simply connected region T conformally onto the interior of the unit circle; how this mapping behaves on the boundary is another question that requires an analysis of angle. The absolute value (i.e. the modulus) of the mapping function in Osgood's 1900 paper approaches the number one as required, but in order to guarantee a continuous mapping on the boundary, one has to show that the argument is also well-behaved. This topic of behavior at the boundary came to the fore once the Riemann mapping theorem was proven. Osgood observed

---
[63] See Osgood, 1903b.
[64] See Birkhoff, 1938, p. 293.
[65] See Osgood and Taylor, 1913.

4. Osgood's Proof of the Riemann Mapping Theorem

that the problem of behavior at the boundary was solved by Schwarz for analytic boundaries, with Picard dealing with the case of the vertices; and it was solved by Painlevé for arcs of boundaries that are convex regular curves.[66] Osgood and Taylor solved the problem for the case of boundary points accessible by a continuous curve. Their result is therefore a nice completion to the 1900 paper, answering definitively the question of behavior at the boundary for the most general region. This is a deep result obtained by an impressive method, but it is not often referred to in the literature.

Osgood's proof of the existence of the Green's function followed methods of Harnack and Poincaré to a great extent. In light of this, one might wonder if Osgood's contribution was small, if he "merely" had to do a bit of cleaning up. As evidence to the contrary, and in addition to Birkhoff's view, consider Ahlfors' 1951 opinion: "Osgood's proof is very remarkable, because it is so clear and concise and does not leave any room for doubt. It is based on an idea of Poincaré, but Osgood deserves full credit for making the idea work." Indeed, Poincaré did not himself produce a full proof of the Riemann mapping theorem until 1908.[67] In this paper, Poincaré himself gave credit to Osgood for the proof of the Riemann mapping theorem, pointed out a difference between his work and Osgood's and reproduced Osgood's method of proof. Poincaré also acknowledged Harnack's contributions. Ahlfors credited Paul Koebe for taking a major step by proving a general uniformization theorem in 1907. He paraphrased Koebe's result in this manner: "every simply connected Riemann surface is conformally equivalent with the sphere, the disk or the plane."[68]

Poincaré's 1883 paper sometimes seems unclear and imprecise. Poincaré shared this view—in his 1908 paper he called the earlier paper his "mémoire primitif" and listed some of the issues it left un-

---

[66] See Osgood and Taylor, 1913, p. 277.

[67] See Poincaré, 1908. Poincaré must have submitted the proof for publication much earlier than 1908, since the paper was actually printed for volume 31 of *Acta Mathematica* in March 1907

[68] See Ahlfors, 1953, p. 6.

clear or treated with insufficient rigor. It was therefore difficult to determine exactly what Poincaré thought he had accomplished in that earlier paper. The 1908 paper provided some clarification. He did not see his 1883 paper as a proof of the Riemann mapping theorem. Thus Osgood used Poincaré's method in a way that Poincaré himself had not precisely envisaged. In particular, the 1908 paper makes clear exactly what Poincaré thought of as Osgood's original contribution. On page one of the paper, Poincaré wrote:

> Mes procédés permettaient bien de démontrer que l'on pouvait faire la représentation conforme de ma surface de Riemann sur une aire intérieure á un cercle; mais on ne voyait pas que ce fût possible sur un cercle; M. Osgood a écarté cette difficulté...[69][70]

He returned to the subject again only two pages later:

> Il s'agit ensuite de montrer que la surface de Riemann peut être représentée sur un cercle; c'est ce qu'a fait Osgood, je reproduis sa démonstration dans le §8, mais en la présentant de manière à faire voir que c'est bien par la fonction z que se fait la répresentation. Je donne ensuite du même théorème une seconde démonstration.[71]

Poincaré thus acknowledged that while his one-to-one function mapped the region he was considering into the unit disk, he had not examined the issue of whether or not the mapping was onto. He recognized that Osgood dealt with this difficulty by showing

---

[69]"My methods permitted me to show that one could carry out the conformal representation of my Riemann surface onto an area interior to a circle; but one did not see that this was possible onto a circle; Mr. Osgood disposed of this difficulty."

[70]See Poincaré, 1908, p. 1.

[71]"Then it is a question of showing that the Riemann surface can be represented on(to) a circle; this is what Osgood did, I reproduce his proof in section 8, but presenting it in such a way as to show that it is indeed by the function z that the representation is accomplished. I then give a second proof of the same theorem." See Poincaré, 1908, p. 3.

4. Osgood's Proof of the Riemann Mapping Theorem

that if a is a point on the boundary of T, if $p_1, p_2,\ldots$ lie in T and have a as their limit point, and if u is the harmonic limit function, then $\lim_{n\to\infty} u(p_n) = 0$, i.e., that the limit function is in fact a Green's function. Poincaré thought enough of Osgood's proof that he included a two-page exposition of the details in his paper.[72]

Osgood's careful attention to mathematical rigor in his proof of the existence of the Green's function was part of the development of the study of normal families. While a bounded infinite set of points always possesses a limit point, this is not always true of an infinite set of functions. It was therefore of interest to know under what conditions an infinite set of functions did admit a limit function. Lebesgue's research on the Dirichlet and Plateau problems, Giulio Ascoli and Arzelà's work on equicontinuity and David Hilbert's research on the Dirichlet problem all involved sequences of functions that admitted a limit function. A family of functions in which every sequence of functions contains a convergent subsequence came to be called a normal family.[73] A more abstract approach to proving the Riemann mapping theorem, using normal families, later became standard in graduate textbooks such as *Complex Analysis* by Ahlfors.

At the dawning of the new century, Osgood's paper on the Riemann mapping theorem represented a triumphal accomplishment in the coming of age of the American mathematical research community. Osgood gave the first proof of the general case of the theorem. As he had done earlier with his term by term integration paper, he again established himself as a participant in the mathematical explorations carried out among eminent European mathematicians such as Poincaré and Harnack. In the process, he combined the geometric imagery of Riemann with the physical intuition of Klein and the mathematical rigor of Weierstrass. He firmly established

---

[72] See Poincaré, 1908, pp. 33–35.
[73] See Montel, 1971, p. 175.

himself, and by extension the American research community, in the midst of the most important mathematical research of the time.

Chapter 5

A Jordan Curve of Positive Area

Osgood's paper entitled "A Jordan Curve of Positive Area" appeared in the fourth volume of the *Transactions of the AMS* in January 1903.[1] Camille Jordan had shown that the area of a rectifiable curve, one of finite length, is zero, but the question of whether a general Jordan curve had zero area was still unanswered. As the title of this paper indicates, Osgood answered the question in the negative, constructing a Jordan curve with positive area. This result may not be as fundamental as those contained in the term by term integration and Green's function papers but it is certainly exciting and intriguing due to its non-intuitive nature. This curve must have been a surprise to many and a horror to some. In his paper, Osgood used Jordan's definition of a Jordan curve:

> The most general continuous plane curve without multiple points may be defined as a set of points which can be referred in a one-to-one manner and continuously to the points of a segment of a right line, inclusive of the extremities of the segment, if the curve is not closed; and

---
[1] See Osgood, 1903a.

to the points of the circumference of a circle, if the curve is closed.

Thus a Jordan curve is the range of a continuous one-to-one mapping of a closed finite interval or a circle into the plane. Careful to place his examination of the subject in the context of current research, Osgood informed the reader that Jordan had already shown that the area of a curve of finite length is always zero. His proposed curve must therefore be infinite in length.[2]

5.1 Motivation for Osgood's Result

The curve of positive area is a natural successor of Osgood's work on the Riemann mapping theorem, which he proved with his 1900 paper establishing the existence of the Green's function. In that paper, he had to take into account the possible complications of simply connected plane regions, and he recognized in particular that such regions could have inaccessible boundary points. Leading up to the Jordan curve paper, he considered other possibilities for the boundary of a simply connected region. In the process, he discovered a closed Jordan curve that bounded a simply connected region with an interior area that was not the same as its exterior area.[3] This construction led him to the Jordan curve of positive area described in his 1903 paper.[4]

A question posed by Camille Jordan provided further motivation for Osgood's paper. Jordan wondered if the area of a parameterized curve might be "indeterminate," and Giuseppe Peano

---

[2] This is most of the first paragraph of Osgood's paper. He cites his source as Jordan's *Cours d'analyse*, vol. 1, 2nd ed., p. 90. He credits the formulation by parametric equations that he will use to Hurwitz, 1898, p. 102. See Osgood, 1903a, p. 107.

[3] Osgood briefly described this construction on pp. 111–112 of his 1903 paper.

[4] See Osgood, 1903a, p. 112.

responded with a reference to his space-filling curve.[5] However, Peano's space-filling curve, possessing multiple points (points where it crosses itself), is not a Jordan curve. Jordan's question brought a gap in the literature to Osgood's attention, and Osgood had an idea about how to fill it. Once again, Osgood actively engaged in the European mathematical dialogue. Osgood was not the only American mathematician interested in curves of positive area. In particular, E.H. Moore at the University of Chicago wrote on the topic of space-filling curves. His paper "On certain crinkly curves," which appeared in the first issue of the *Transactions*, was described by R.C. Archibald as delightfully illuminating the topic of the space-filling curves of Peano and Hilbert.[6]

5.2 Historical Context—What is *area*?

In the latter part of the 19th century and up until the work of Borel and Lebesgue on measure became well known, there was not a consensus among mathematicians about the "right" definition of the concept of measure, even in the one-dimensional case. In two dimensions, the situation was even more complicated. A definition was needed that agreed in some acceptable way with intuitive geometric notions. Yet the developments in point set theory had led to the need to assign numbers for length, area and volume to intuitively difficult sets. This also had direct implications in the hunt for good definitions of the integral, an active topic of research around the turn of the century, as described in Chapter III on Osgood's paper "Non Uniform Convergence and the Integration of Series Term by Term."

---

[5] Osgood gave the references to Jordan's question and Peano's reply as follows: *Mathematische Annalen*, vol. 36, 1890 and *L'Intermédiaire des mathématiciens*, vol. 1, p. 23, question 60, 1894 and vol. 3, p. 30, 1896.

[6] See Archibald, 1938, p. 147.

Mathematicians of the time were well aware of the non-intuitive results that could be found by exploring the logical consequences of proposed definitions. For example, the intuitive definitions of a function that were in use in the 18th and 19th centuries had been the object of much debate, even discord. As they explored the consequences of definitions, mathematicians found examples of more and more non-intuitive features of functions. Dirichlet gave an example of a function which is not given by an analytic expression and is discontinuous everywhere. Weierstrass exhibited a continuous function that is nowhere differentiable. Hermite called such functions "lamentable evil" and Poincaré thought of them as monsters that did not resemble "honest functions which serve some purpose."[7] Thus mathematicians were conscious of the pitfalls inherent in trying to provide a definition for an intuitive concept such as area. The idea that there might be a continuous, one-to-one curve with positive plane measure is certainly non-intuitive, and is surprising to many. Osgood's discovery of this curve constituted a link in the broader exploration of non-intuitive consequences of definitions of area, exterior area and interior area.

What were some of the ideas of area, length and volume of point sets in use around the time Osgood wrote this paper? As previously noted, some of the earliest were due to Axel Harnack and Georg Cantor in the early 1880s.[8] In 1885, Harnack published the paper Osgood used as a basis for the idea of content in his term by term integration paper.[9] Recall that Harnack defined his content of a subset, M, of the real line as the infimum of the lengths of coverings of the set by a finite number of non-overlapping closed intervals—an exterior measure. Harnack did not extend his notion of content to higher dimensions using coverings by a finite number of non-overlapping closed squares or rectangles. Instead he adapted his exhaustion of the complement procedure, and removed

---

[7] See Kline, 1972, p. 973.
[8] See Cajori, 1924, pp. 404–405.
[9] See Harnack, 1885.

5. A Jordan Curve of Positive Area 127

n-dimensional spheres of radius $1/k$ from an n-dimensional interval containing the set M to be measured.

Osgood broke with his tradition of using Harnack's content when he wrote this Jordan curve paper; he preferred instead to use an exterior area defined by Giuseppe Peano. To find the Peano exterior area of a set, μ_P, specifically in Osgood's case of a curve C, one covered the plane with non-overlapping squares of side length 2^{-n}, then formed the sum, s_n, of the areas of all the squares containing a point of the curve in their interior or on their boundary. The $\{s_n\}$ are a decreasing sequence whose limit was called the exterior area of the curve, i.e. $\mu_P(C) = \lim_{n\to\infty} s_n$.[10] Because any covering of P by a finite number of non-overlapping closed squares or rectangles is also a covering of the closure of P, given any open subset P of $\mathbb{R}^2$, $\mu_P(P) = \mu_P(\overline{P})$.

In 1893, Jordan published a definition of measure in his *Cours d'analyse de l'École polytechnique*.[11] This definition became commonly accepted until Borel/Lebesgue measure eventually supplanted it. Jordan divided a plane interval, for example, into a finite number of squares of side length s. He defined S to be the sum of the areas of all squares whose points all belong to a set P, then let $S + S'$ be the sum of the areas of the squares that contain any point of P. Then as s approaches zero, S converges to a limit, I, called the interior area, and $S + S'$ converges to a limit, A, called the exterior area. If I equaled A, then P was said to be *measurable*.[12] Using Jordan's measure, note that the set of rational numbers in the unit interval is not measurable since $I = 0$ while $A = 1$. The difficulty is again caused by the requirement of a finite covering. Although Osgood cited Peano as his source of a definition of exterior area, by 1903 he was also familiar with the edition of Jordan's *Cours d'analyse* containing Jordan's defi-

---
[10] See Peano, 1887, p. 156.
[11] In Schoenflies, 1899, p. 91, there is also a reference to an 1892 paper in which Jordan published his definition of measure—*Journ. de Math.* (4) 8: 77, 1892. See Jordan, 1893.
[12] See Bliss, 1917–1918, p. 24 and Schoenflies, 1899, p. 92.

nitions.[13] Using these definitions, Osgood's curve of positive area is not measurable—it has interior Jordan measure 0 and positive exterior Jordan measure.

Following Borel's ground-breaking work, Lebesgue's exterior measure was later defined via the infimum of the areas of coverings of an arbitrary subset, P, of the plane for example, by a *countable* collection of closed rectangles. In Jordan's tradition, a suitable interior measure can be defined and then a *measurable* set can be defined as one for which the exterior measure equals the interior measure. The exterior Lebesgue measure of the rationals in the unit interval is 0, as is the interior measure, because countable coverings are permitted. This major breakthrough eliminated the discrepancies caused by the use of finite coverings in the definitions of Harnack, Peano and Jordan.

For a more current understanding of Osgood's construction of a Jordan curve of positive area, it is important that in the case of a closed, bounded subset, F, of the plane, $\mu_P(F) = \mu_L(F)$. Osgood's curve is a closed bounded subset of the plane and hence its Lebesgue measure is positive. The fact that his curve has positive area is not attributable, therefore, to his use of Peano's exterior area, which is based on finite coverings. To see the equivalence of Peano and Lebesgue measure in this case, consider the set, P, of coverings of F by a countable collection of closed rectangles R. Let Σ be the set of coverings of F by a finite collection of closed squares S. Since $\Sigma \subset P$, taken over all coverings,

$$\inf | \cup R | \leq \inf | \cup S |$$

which by definition means that

$$\mu_L(F) \leq \mu_P(F).$$

---

[13] See Osgood, 1903a, p. 107.

Given $\varepsilon > 0$, there is a covering $\wp$ of F by open rectangles such that
$$\mu_L^{\text{ext}}(\wp) \leq \mu_L^{\text{ext}}(F) + \varepsilon.$$
F is compact so there is a finite subcover Δ of $\wp$. Then
$$\mu_L^{\text{ext}}(\Delta) \leq \mu_L^{\text{ext}}(F) + \varepsilon.$$
But
$$\mu_P(F) \leq \mu_L^{\text{ext}}(\Delta)$$
since the Peano area is the infimum of the area of the closures of the rectangles in finite coverings such as Δ, which is equal to the infimum of the area of the open rectangles. Thus
$$\mu_P(F) \leq \mu_L^{\text{ext}}(F) + \varepsilon = \mu_L(F) + \varepsilon.$$

5.3 Osgood's Construction of the Curve

To construct Osgood's curve, start with an open square of side length 1. Construct the first stage leading to the curve as in Figure 5.1. This resembles Osgood's figure as it appeared, in color plates, in the *Transactions*. Osgood's clever aid to visualizing the curve at each stage was to think of the yellow regions as dikes, the blue regions as canals of water, and the red line segments as pathways that separate water from land. At each stage, a pathway from A to B is thought of as a walk with a canal always on the right side, taking each red pathway segment as it comes. At stage two, repeat stage one in each of the 3^2 white regions as in Figure 5.2. At stage three, repeat stage one in each of the 3^4 white regions of stage two; then continue this process. Let s_n be the area (the usual length times width) of the canals at stage n; this is equal to the area of the dikes at stage n.

Osgood assigned width to the canals and dikes in such a way that s_n converged to a positive number $\lambda < 1/2$ as follows. Let

$\varepsilon_1 + \varepsilon_2 + \varepsilon_3 + \cdots$ be a convergent series of positive numbers whose sum is λ, where $\lambda < 1/2$. At stage one in the construction, choose the width of the canals such that the total area of the stage one canals will be ε_1. For example, if w is the width of the canals, the length of the canals at stage one is $2 - w$. Thus choose w so that $\varepsilon_1 = w(2 - w)$. Use the same width for the dikes so their total area is also ε_1. At stage n, chose the width of the canals such that the area of the canals is ε_n and do the same for the area of the dikes. It then follows that the area of the canals and the dikes in the construction is $2 \lim_{n \to \infty} s_n = 2\lambda < 1$.

5. A Jordan Curve of Positive Area

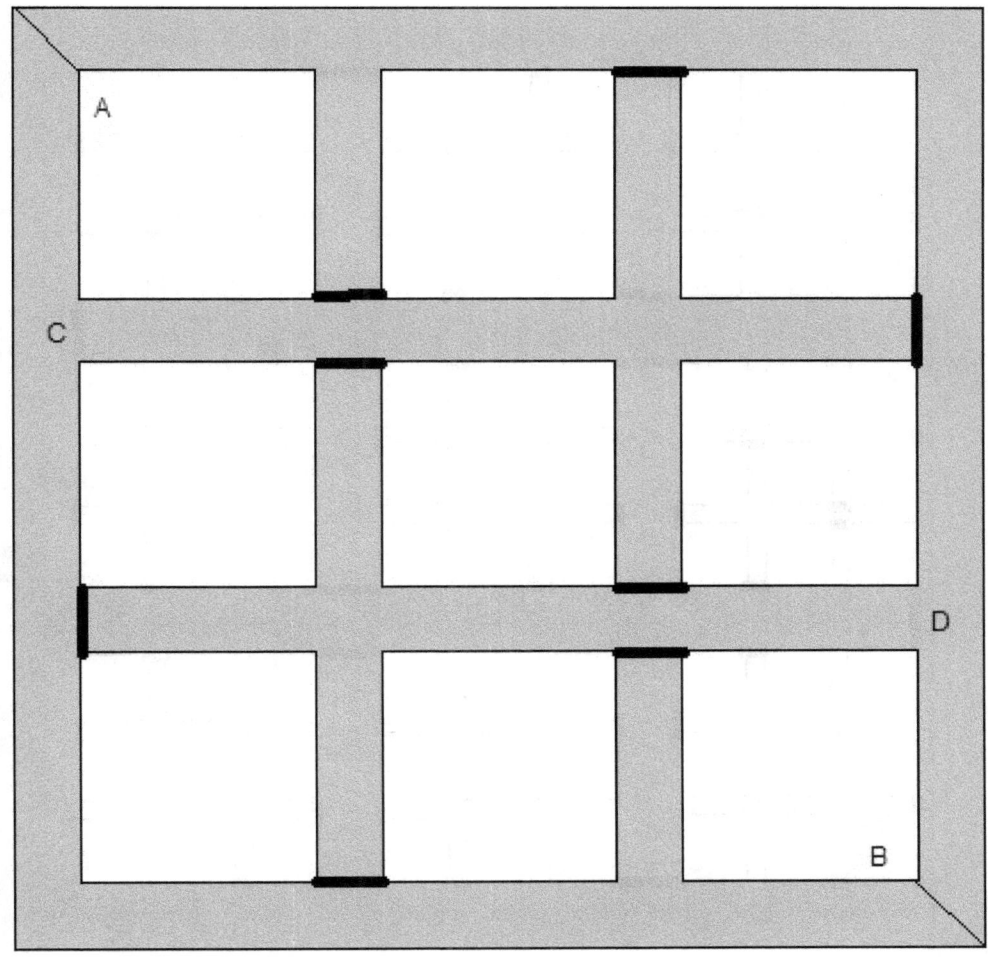

Figure 5.1: Osgood's curve construction—stage one

- *A* is the corner where the curve begins.
- *B* is the corner where the curve ends.
- *C* marks Osgood's blue canals at stage one.
- *D* marks Osgood's yellow dikes at stage one.
- The thick black lines represent Osgood's stage one red pathways separating water from land.

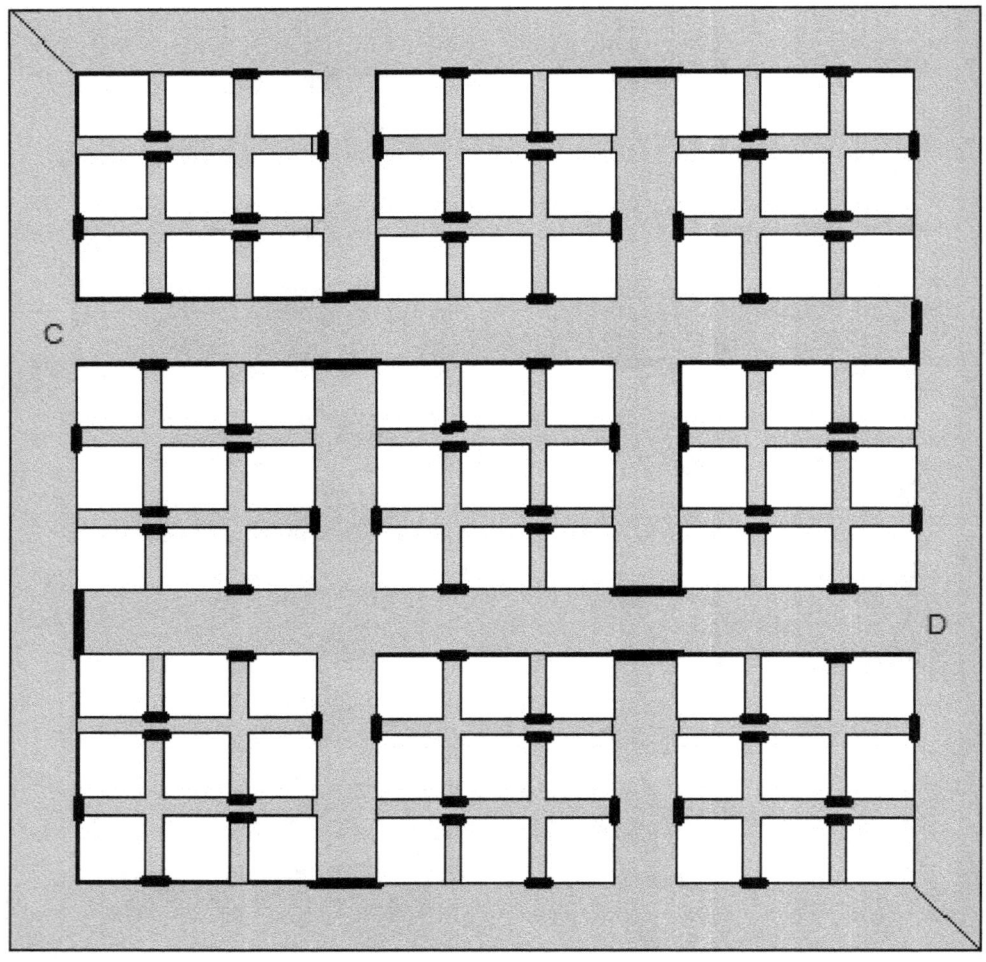

Figure 5.2: Osgood's curve construction—stage two

- C represents the canals at stage two
- D represents the dikes at stage two
- the shorter thick black lines are the pathways constructed in stage two

5. A Jordan Curve of Positive Area

At the limit, the canals, as added at each stage, form a union of a countable number of non-overlapping open $\mathbb{R}^2$ intervals, together with a countable number of boundary segments where a new segment of canal is joined to one previously constructed. Call this set of canal segments and corresponding boundary segments W. The Lebesgue measure of the set of boundary points is 0. The union of the canal segments is an open set, which is Lebesgue measurable; the same is true of the dikes, L. Then $\mu_L(L) = \mu_L(W) = \lambda$. The canal/dike framework, $L \cup W$, can be viewed as an open set, although whether Osgood himself viewed it in such a light is unknown.[14] Thus its complement in the closed square is closed, hence measurable. If this complement of $L \cup W$ is the curve, C, then the $\mathbb{R}^2$ Lebesgue measure of C is $1 - 2\lambda$, which is positive since $\lambda < 1/2$. C is closed and bounded so the Peano exterior area is also $1 - 2\lambda$.

The union of the red pathways, together with their limit points, forms a set which Osgood represented as a Jordan curve from point A to point B (see Figure 5.1) by showing its correspondence to a line segment $0 \leq t \leq 1$ where A corresponds to $t = 0$ and B to $t = 1$.

In detail, at stage one assign values of t in $(0, 1/17)$ to the path along the canals before the first red segment. The values in $[1/17, 2/17]$ are assigned to the first red segment; the values in $(2/17, 3/17)$ to the next segment of path along the canals, the values in $[3/17, 4/17]$ to the next red segment. Continue in this way until the last segment of path along canals is assigned values in $(16/17, 1)$. At the end of stage one, eight red segments have been constructed. At stage two, there are eight new red segments in each of the nine squares. In the upper leftmost square, the path along the canals to the first red segment is assigned the values in $(0, 1/17^2)$, the first red segment is assigned values in $[1/17^2], 2/17^2]$, the next path along the canals is assigned values in $(2/17^2, 3/17^2)$. This continues until the last segment of path along the canals before arriving at

---

[14] At any stage n, in fact, the canals, or respectively the dikes, form an open set.

the first red segment constructed in stage one is assigned values in $(16/17^2, 1/17)$. Repeat the process in the remaining eight squares. Generally, at stage n, new red segments in the upper leftmost square are assigned values in

$$\left[\frac{2k-1}{17^n}, \frac{2k}{17^n} \right],$$

for $k = 1, 2, \ldots, 8$. The remaining squares are assigned values in the same fashion using appropriate values of k. To complete the curve, Osgood adjoined to red segments all limit points of the red segments that were not already part of the construction.

Referring to the red pathways in a footnote, Osgood stated that it "is readily seen that the content of the set of points of condensation of the extremities of their segments is null."[15] Assuming Osgood meant that all the area of the curve was contained in the red segments and none in the limit points, he was mistaken. The red segments alone form a countable collection of non-overlapping closed $\mathbb{R}^1$ intervals, each with two dimensional Lebesgue measure 0. So the collection of red segments has Lebesgue measure 0. Therefore, the limit points must account for the entire exterior area of the curve. But this reasoning with Lebesgue measure was not available to Osgood. Indeed, to the contrary, he knew that it was possible to form a set of positive Peano measure that is the union of a countably infinite number of sets of Peano content zero. It was, therefore, not entirely unreasonable for him to assume that the area of the set of red segments could be positive.

Osgood showed in the paper that C is a Jordan curve and he ensured that the combined area of the canals and dikes, $L \cup W$, is $2\lambda < 1$. He assumed then that the curve is the complement of $L \cup W$ in the square so that the area of the curve is $1 - 2\lambda$, a positive number. Osgood's assumption that the curve is the complement of $L \cup W$ is reasonable, but he offered no argument for it. To examine

---

[15]See Osgood, 1903a, p. 110.

the nature of the limit points which make up the positive content of the curve C, consider a candidate point P that is not in $L \cup W$. At any stage n, P must be in a white square, being neither on a dike, nor in a canal, nor on a previously constructed red segment. The diameter of the white square at stage n is less than $\sqrt{2}/3^n$ so the distance from P to a red segment at stage n is less than $\sqrt{2}/3^n$. Thus P is in a nested sequence of white squares and is the limit of two sequences of the red corners of the squares, which means that P is on the curve C. This type of nested interval reasoning had been used by Osgood six years earlier in his term by term integration paper—he presumably could have supplied such an argument had he felt it necessary.

For the parametrization of the curve to be one-to-one, it was necessary to show that, given a point P, there is a unique value of t such that the location of P is given by the parametric function

$$\vec{P}(t) = (\phi(t), \psi(t)).$$

Osgood argued as follows. Given $\eta > 0$ and a point P, it is possible to find an N and two red segments such that the red segments have been assigned values of t in

$$t_N = \left[\frac{2k-1}{17^N}, \frac{2k}{17^N}\right]$$

and

$$T_N = \left[\frac{2k+1}{17^N}, \frac{2k+2}{17^N}\right]$$

for an appropriate choice of k and

$$\frac{2k+1-2k}{17^N} = \frac{1}{17^N} < \eta.$$

For all $n \geq N$, the value of k may vary but the distance between the two corresponding red segments of that stage remains less than η. In fact, the right endpoints of the t_N form an increasing sequence while the left endpoints of the T_N decrease. Since η is arbitrary,

$$\sup_{n \geq N} t_n = \inf_{n \geq N} T_n$$

Assign this common limit value of t_p to the point P so its location is given by $\vec{P}(t_p)$. Given any distinct limit point P_1, at some stage m, P and P_1 will be in different white squares. At that stage, the t values assigned to the red segments adjoining the white square of P_1 will be different from those of P (some land or water will intervene) and thus the limit value will be different.

To show that the curve C is continuous, Osgood considered two cases. If t is interior to one of the intervals corresponding to the interior of a red segment, then $\vec{P}(t)$ is interior to a red segment and $\vec{P}$ is clearly continuous there. If t is not such a point, Osgood justified the continuity of $\vec{P}$ at t by stating that "a neighborhood of the corresponding point (x, y) may be marked off such that the parametric values of t corresponding to the points of the curve lying in it will differ arbitrarily little from the value of t in question, and will completely fill a certain neighborhood of this value."[16] Here Osgood appealed, in somewhat sketchy fashion, to an understanding that given such a t and a small neighborhood V about $\vec{P}(t) = (x, y)$, one can see that $\vec{P}^{-1}(V)$ is a small interval about t because of the way the curve spirals around such a point infinitely many times.

As noted previously, Osgood was motivated to find a curve of positive measure by finding a simply connected region of the plane whose exterior area is not the same as its interior area. The canals in Osgood's construction are also such a region. To show this, use a Peano-like interior area of a set P defined by forming the sum, S_n, of all squares containing only points of the curve in their interior or on their boundary. Then, $\{S_n\}$ forms an increasing sequence whose limit is the interior area. The canals were constructed in such a way that
$$\mu_P^{\text{int}}(W) = \mu_L(W) = \lambda,$$

---

[16] See Osgood, 1903a, p. 111.

while a calculation of the exterior area gives

$$\begin{aligned}\mu_P^{\text{ext}}(W) &= \mu_P^{\text{ext}} P(\overline{W}) \\ &= \mu_P^{\text{ext}}(W \cup C) \\ &= \mu_P^{\text{ext}} L(W \cup C) \\ &= \mu_P^{\text{ext}} L(W) + \mu_L^{\text{ext}}(C) \\ &= \lambda + (1 - 2\lambda) = 1 - \lambda.\end{aligned}$$

5.4 Enduring Interest in Osgood's Result

Osgood's curve, and others of positive measure subsequently constructed, have had a lasting impact. Osgood's colleague, J.L. Walsh, proved in 1926 that if γ is an arc in the complex plane then every continuous function on γ is approximable there by polynomials in z.[17] In 1955, John Wermer used the existence of an arc of positive two dimensional Lebesgue measure to prove that the analogous result is not true in $\mathbf{C}^3$.[18] In Wermer's 1971 text, *Banach Algebras and Several Complex Variables*, he noted that Osgood's curve was the first example of an arc of positive plane measure.[19] Andrew Browder, in his text *Introduction to Function Algebras*, discussed Wermer's result and again mentioned Osgood as the first to demonstrate the existence of such an arc.[20]

Osgood's curve is also an early example of a fractal. Benoit Mandelbrot explored the fact that fractals are "dimensionally discordant" by using two types of dimension, the Hausdorff dimension and the usual topological dimension by which any set in Euclidean space is assigned a number which "on intuitive and formal grounds strongly deserves to be called its dimension." *Hausdorff dimension* (also called *Hausdorff Besicovitch dimension*) is a non-negative real

---
[17] Walsh's result was announced in the *Bulletin of the AMS*, 32: 35, 1926.
[18] See Wermer, 1955.
[19] See Wermer, 1971.
[20] See Browder, 1969, pp. 200–201.

number associated with a metric space that generalizes the concept of topological dimension (the n in $\mathbb{R}^n$) to other metrics.

Mandelbrot defined a fractal to be "a set for which the Hausdorff Besicovitch dimension strictly exceeds the topological dimension."[21] He then argued that, while Osgood's curve has topological dimension one, it has Hausdorff dimension two.[22]

Mandelbrot showed his enthusiasm for curves of positive area and surfaces of positive volume, exclaiming:

> What a mad combination of contradictory features! Have we not finally come to mathematical monsters without conceivable utility to the natural philosopher? Again, the answer is emphatically to the negative. While believing they were fleeing Nature, two famous pure mathematicians unknowingly prepared the precise tool I need to grasp (among others) the geometry of ... flesh.[23]

Osgood and Lebesgue, who also exhibited a curve of positive area, were his two famous pure mathematicians, and his tool to grasp the geometry of flesh was a model of the human vascular system. Mandelbrot went on to call curves of positive area "hard to swallow" and wrote that "after Lebesgue and Osgood showed that swallow them we must, they came to supersede the Peano curve as supreme monsters."[24] But he intended to show that they were not actually monsters. Mandelbrot believed that a variant of Osgood's curve with topological dimension two and Hausdorff dimension three, far from being a monster, roughly modeled the vascular system of the human body. The canals and dikes became veins and arteries and

---

[21] See Mandelbrot, 1983, p. 15.
[22] See Mandelbrot, 1983, p. 147.
[23] See Mandelbrot, 1983, p. 147.
[24] See Mandelbrot, 1983, p. 148. This almost certainly referred to Poincaré's lament about functions that belong in a gallery of monsters because they serve no honest purpose. See also Mandelbrot, 1983, pp. 3 and 9.

5. A Jordan Curve of Positive Area 139

the boundary corresponded to nonvascular tissue, with the assumption that every point of nonvascular tissue was infinitely near a vein or an artery.[25]

Mandelbrot appeared to disparage Osgood's motivation for exhibiting a curve of positive area, referring to Osgood's "fanciful way of making a contrived construction easier to follow" and stating that:

> No one cares that [such curves/surfaces] first arose in a contrived mathematical flight from common sense. I have shown that they are intuitively unavoidable, that *Lebesgue/Osgood fractal monsters are the very substance of our flesh!*[26]

These comments, combined with the above reference to "pure mathematicians unknowingly" preparing the tool Mandelbrot needed, seem to imply that he considered Osgood a pure mathematician with no respect for or interest in applications. This might give a reader the wrong impression of Osgood—nothing could be farther from the truth. Osgood's interest in potential theory, as described in Chapter 4 on his proof of the Riemann mapping theorem is one piece of evidence that, in addition to his pure mathematics work, Osgood had a keen interest in the applications of mathematics. Chapter 6, in which some of Osgood's views about teaching and the undergraduate curriculum are described, will also show his respect for applications. In particular, he believed that the foundation for the study of calculus was physics; that its purpose in undergraduate education was to give students an understanding of the physical world.

Osgood's demonstration of a Jordan curve of positive area was a delightful example of how relying on intuition can be unreliable in

---
[25] See Mandelbrot, 1983, p. 149.
[26] See Mandelbrot, 1983, pp. 149, 150.

mathematics, and thus again confirmed Osgood as a proponent of rigor. Moreover, the manner in which it was counterintuitive made it an early example of a fractal, a topic that would be of popular interest later in the century. Osgood's result again placed him in the midst of the European dialogue in mathematics, answering a question posed by Camille Jordan, and working in the same arena as Giuseppe Peano and Henri Lebesgue.

Chapter 6

Osgood's Teaching—Textbooks, Students and Approaches

Previous chapters have indicated how William Fogg Osgood's research helped bring the United States into the international mathematical arena. Osgood and colleague Maxime Bôcher also played a part in shaping modern collegiate mathematics education. Harvard in the mid-19th century had Benjamin Peirce, but Peirce did not produce a next generation of researchers. The state of undergraduate mathematics education at Harvard had started to improve with the introduction of continental European mathematics texts, but this had not yet resulted in strong graduate programs or in a research community. Parshall and Rowe identified a key event in the emergence of the American mathematical research community as the 1876 installation of J.J. Sylvester at The Johns Hopkins.[1] During this period, Charles Eliot and James Mills Peirce also created a favorable climate for research at Harvard. Osgood and Bôcher responded to that favorable climate; contributed to Harvard's part in putting American research mathematics on the map; and helped

---
[1] See Parshall and Rowe, 1994.

ensure the continued growth of the newly emerged mathematical research community.

Parshall and Rowe described the period from 1900 to 1933 as one of consolidation and growth of the American mathematical research community. Throughout much of this period, Osgood made sure that new generations of Harvard students were well trained in mathematics, especially at the undergraduate level but also at the graduate level to some extent, thereby making some contribution to the growth of the new research community. Generations of students used edition after edition of Osgood's elementary textbooks, even after his retirement from Harvard in 1933. His two-volume treatise *Lehrbuch der Funktionentheorie*[2] guided mathematicians and their upper-level students through a rigorous treatment of complex functions in one and several variables. He participated with fellow mathematicians like E.H. Moore in the debate on what the undergraduate mathematics curriculum should include[3]; and his AMS presidential address presented his vision for the teaching of calculus in colleges and technical schools.[4] Osgood was one of three American commissioners, with D.E. Smith and J.W.A. Young, of the International Commission on the Teaching of Mathematics, created at the fourth International Congress of Mathematicians in Rome in 1908 with Felix Klein as President.

Since Osgood was so deeply involved in issues related to education and high-level research, it seems curious that he served as thesis advisor for only four doctoral students. Maxime Bôcher provided a complement to Osgood in this respect, producing 17 Harvard Ph.D.s in mathematics during his short career. The teaching styles of the two men complemented each other in various ways. J.L. Coolidge compared them in his essay on the history of mathematics at Harvard:

---

[2] See Osgood, 1907a.
[3] See Maltbie, 1900.
[4] See Osgood, 1907b.

6. Osgood's Teaching

Let us return to Osgood and Bôcher, and inquire what were the effects of their addition to the Division of Mathematics [at Harvard]. To begin with, two teachers were added of equal but opposite excellence. Osgood acquired skill in teaching by the same process that brought him scientific eminence: conscientious effort and high ideas. He found out by experiment what were the most important things to teach, and what was the best way to teach them. Nothing was left to chance or the inspiration of the moment. When he finished there were no loose ends. Bôcher never made an appreciable effort to be clear or interesting. His teaching *was* clear, because his mental processes were like crystal; his teaching *was* interesting because he cared about interesting things. Yet the principal contribution of these two men was to establish traditions and standards as a guide to the future.[5]

Bôcher also participated in the American efforts of the International Commission on the Teaching of Mathematics, serving as chairman of the American Committee on Graduate Work in Universities.

From the international level to the local at Harvard, from the most advanced graduate students to the beginning freshmen, Osgood and Bôcher were among the shapers of modern mathematics education in the United States. They exemplify the fact that Felix Klein's "dedication to and interest in mathematics education ultimately proved to be a characteristic feature within the early American mathematical community."[6] A full analysis of the shaping of collegiate mathematics education in the United States around the turn of the century would include the influences of E.H. Moore at University of Chicago, James Pierpont at Yale and others. This chapter examines the particular part of William Fogg Osgood at Harvard in the shaping of collegiate mathematics education, high-

---
[5] See Coolidge, 1930, pp. 250–251.
[6] See Parshall and Rowe, 1994, p. 191.

lighting also many of the contributions of Maxime Bôcher that complemented Osgood's efforts.

6.1 Textbooks, Calculus Reform and the Use of Infinitesimals

Osgood's textbooks had an impact on many students, indeed on nearly every student who took a mathematics course at Harvard until well after his retirement in 1933. His first textbook was a 71-page pamphlet published in 1897 and entitled *Introduction to Infinite Series*. It could be obtained from the Publication Agent of Harvard University for the price of 75 cents. In his review, Canadian mathematician James Harkness marveled:

> That it is possible at once to interest the reader, to make no sacrifice of thoroughness, and to arrange the material in organic connection with the other parts of mathematics is proved by Professor Osgood's short pamphlet on infinite series.[7]

Harkness also remarked on Osgood's practice of never giving a proof of a theorem until the theorem actually required a proof. This marked the beginning of an idea that Osgood carried throughout his career—rigor was a relative matter that depended on the level of the students. This meant that Osgood considered a curriculum rigorous if it was presented with all the logic that the students were capable of handling at their level.

---
[7] See Harkness, 1898, p. 277.

6.1.1 Calculus reform

Osgood's early years at Harvard were a time of transformation in patterns and standards of undergraduate and graduate education. Calculus reform was a part of this transformation. George M. Rosenstein described the situation:

> Between 1885 and 1907, the number of members of the AMS doubled to 568 and a single section became four. Moreover, the presidents of the organization were young. Of the first ten (through 1910), only Van Amringe, McClintock and indefatigable Newcomb were over fifty when they presided. Half of them had studied in Germany. In an age that cherished "progress," traditionalists would have been hard pressed to stop the rush of these enthusiastic students of brilliant German teachers to reform the teaching of the calculus.
>
> The Bulletin of the Society continued to be filled with reports on teaching mathematics at all levels and on teaching calculus in particular. Osgood's presidential address in 1907 was called "The Calculus in Our Colleges and Technical Schools." Importantly, calculus books were reviewed critically in the Bulletin. Old publications were pushed out and new "modern" books took their place.[8]

Osgood's first major textbook, *A First Course in the Differential and Integral Calculus*, was one such "modern" calculus book, and it came out in the same year as his 1907 presidential address on the calculus. This timing was duly noted by Charles N. Haskins of the University of Illinois who reviewed the 1909 revised edition of the textbook:

---
[8] See Rosenstein, 1989, p. 97.

> Professor Osgood in his presidential address before the American Mathematical Society has discussed and illustrated the principles which his experience has led him to consider should govern the teaching of the calculus. In the present text he gives us the detailed application of those principles to the difficult pedagogical problems which confront the instructor in the first course in this subject.[9]

Haskins' review leads to some questions. What were these principles articulated by Osgood in his presidential address? How were these applied in his textbook? Why was it modern?

Osgood firmly believed that the foundation for the study of calculus was physics. He thus began his address by calling attention to the history of the subject:

> The history of the race is frequently suggestive, in educational matters, of a wise course for the training of the individual. If we turn to the problems to which the calculus owes its origin, we find that not merely, not even primarily, geometry, but every other branch of mathematical physics—astronomy, mechanics, hydrodynamics, elasticity, gravitation, and later electricity and magnetism—in its fundamental concepts and basal laws contributed to its development and that the new science became the direct product of these influences.[10]

So Osgood's ideal calculus class would study applications to physics. The fact that the historical development of the subject substantially paralleled the exploration of physical concepts appealed to Osgood, as evidenced by the passage above. Moreover, he believed that the basic idea students should take away from a calculus course was an

---
[9]See Haskins, 1909, p. 457.
[10]See Osgood, 1907b, p. 449.

6. Osgood's Teaching

understanding of the physical world. He felt that textbooks did not emphasize this understanding sufficiently, and he underscored this in closing through a series of rhetorical questions:

> And yet, with all of this that is so good and sound, if you open one of our text-books on the calculus and ask: What *is* the calculus? What will abide after the formulas are forgotten? What is the soul and the spirit of this great science, as conceived by the man whose work in life does not lie within the field of mathematics? I can't help feeling that the answer does not ring clear: The calculus is the greatest aid we have to the appreciation of physical truth in the broadest sense of the word.[11]

Osgood was not remaking an obvious point. His view was an affirmation of the break, initiated by Farrar and others in the early 19th century, with the more geometric tradition that the mathematics curriculum in the United States originally shared with its British antecedents. For Osgood, calculus should not be used solely to solve geometric problems in the tradition of Euclid; it should be used to solve problems arising in the physical world. In his own research, Osgood was a pure mathematician; in his view of teaching calculus, he was often a mathematical physicist.

How did this view manifest itself in *A First Course in the Differential and Integral Calculus*? According to Haskins:

> Professor Osgood has long emphasized by precept and example the importance of developing new mathematical concepts in the student's mind by means of problems, i.e., of causing the new mathematical idea to appear as a necessary element for the solution of a definite geometrical or physical problem. His chapter on the definite integral

---
[11] See Osgood, 1907b, p. 467.

as the limit of the sum is an example of this method, as well as of gradual development of precision in concept and demonstration conformably to the student's advance in assimilative power.[12]

In particular, Haskins mentioned the problem of fluid pressure on a vertical surface with curved boundaries as Osgood's means of introducing upper and lower integrals. The textbook also contains a chapter on mechanics and has numerous physical application problems scattered throughout. Haskins made a point of the fact that Osgood "has by no means neglected geometry, as his chapters on the cycloid, on curvature and evolutes, on envelopes, on partial differentiation, and on definite integrals, bear witness."[13]

That Haskins found it important to mention that Osgood did not neglect geometry, particularly since Osgood had made the point that applications to the physical world were the main motivation for a student to study calculus, merits an explanation. Haskins' emphasis reflected Osgood's position with respect to the Perry movement. John Perry was Professor of Mechanics and Mathematics of the Royal College of Science in London and chairman of the Board of Examiners there for engineering and other technical subjects. E.H. Moore, in his AMS presidential address, talked a great deal about Perry and his English movement.[14] In general terms, a main purpose of the Perry movement was to promote an elementary mathematics curriculum (through lower undergraduate courses) that contained only practical and applied mathematics. Perry himself wrote textbooks in practical mathematics, applied mechanics and calculus for engineers. Moore explained it:

> One important purpose of the English agitation is to relieve the English secondary school teachers from the bur-

---

[12] See Haskins, 1909, p. 460.
[13] See Haskins, 1909, p. 462.
[14] See Moore, 1903.

6. Osgood's Teaching

den of a too precise examination system,... in particular, to relieve them from the need of retaining Euclid as the sole authority in geometry, at any rate with respect to the sequence of propositions. Similar efforts made in England about thirty years ago were unsuccessful.[15]

So to a certain extent English lower level mathematics instruction was still in a pre-17th century state, and Perry wanted to change that. Moore agreed with Perry's basic idea—that elementary mathematics instruction, whether in secondary schools or colleges, needed to cover more "modern" 17th-century mathematics:

The troublesome problem of the closer relation of pure mathematics to its applications: can it not be solved by indirection, in that through the whole course of elementary mathematics, including the introduction to the calculus, there be recognized in the organization of the curriculum no distinction between the various branches of pure mathematics and likewise no distinction between pure mathematics and its principal applications? Further, from the standpoint of pure mathematics: will not the twentieth century find it possible to give to young students during their impressionable years in thoroughly concrete and captivating form the wonderful new notions of the seventeenth century?[16]

Moore thus acknowledged that reform was also needed in the curricula of colleges in the United States, believing in particular that calculus should be taught along with its practical applications, with no distinction made between the theory and those applications. But by 1909, Haskins felt that the trend toward practical applications had gone too far:

---

[15] See Moore, 1903, p. 410.
[16] See Moore 1903, p. 424.

> The earlier American texts in the calculus have in general confined their applications to the field of geometry. The adherents of the "Perry movement" have gone to the other extreme and have produced a flood of problems taken from engineering practice. They have, however, too often forgotten that the student for whom the problems are designed has not yet acquired the technical knowledge necessary for an appreciation of their meaning and importance.[17]

Osgood's textbook thus struck a balance; it provided the applications to physical phenomena that brought instruction into the modern age, but neglected neither geometric problems that could provide an introductory learning experience nor pure mathematical foundations.

Osgood expressed his own early view of the Perry movement in a review of George A. Gibson's *An Elementary Treatise on the Calculus*.[18] Although Osgood appeared to believe that Perry's problems involved inappropriate technical difficulties, he was generally favorable to the idea behind the movement. He declared:

> We are in hearty sympathy with Professor Perry as regards the objects which he has in view, for we believe that calculus should be brought home to the student and that he should through many and varied problems be brought to feel that calculus stands in vital relation to the phenomena of everyday life. No more fatal criticism on a course in calculus can be made than that which is contained in the remark of the student who says that he does not see what calculus is for.[19]

---
[17] See Haskins, 1909, p. 462.
[18] See Gibson, 1901.
[19] See Osgood, 1902, p. 251.

Osgood's own research was clearly influenced by the research of Axel Harnack, as described in earlier chapters. In the review of Gibson's textbook, Osgood demonstrated that he was also influenced by Harnack's calculus textbook, *Elements of the Differential and Integral Calculus*. This text was first published in German in Leipzig in 1881 and later translated into English. In Osgood's view, it "gave the first systematic presentation in the English language of the leading principles of modern analysis in their relation to the foundations of the infinitesimal calculus"[20]. This praise was tempered by the sentence which followed it:[21]

> While not wholly free from errors, and sometimes difficult to read, owing to inadequate exposition of details, the book is nevertheless conceived in the spirit of modern mathematics and it lays stress on those principles of analysis which are essential for a rigorous development of the calculus.

6.1.2 Infinitesimals

In the search to determine what modern calculus instruction in the United States should look like, Osgood was also in the middle of the debate on use of infinitesimals. Osgood was a firm believer in using infinitesimals in teaching calculus, finding them useful in teaching the applications to physics that he found so important. Still, Osgood did not use infinitesimals in his research, where he worked with all the rigor of the Weierstrassian ε's and δ's. He believed, however, that the use of infinitesimals could provide a level of rigor appropriate to calculus students, and his calculus textbooks clearly reflect this belief. Both *A First Course in the Differential*

---
[20] See Osgood, 1902, p. 248.
[21] See Osgood, 1902, p. 248.

Figure 6.1: Presidents of the American Mathematical Society

and Integral Calculus[22] and the much later revision *Introduction to Calculus*[23] made use of infinitesimals.

Saunders Mac Lane served as a Benjamin Peirce instructor at Harvard from 1934 to 1936, arriving the year after Osgood retired. He recalled that at the time it was proper for the faculty at a major university to produce its own calculus texts, and Osgood wrote them at Harvard. Mac Lane used *Introduction to Calculus* and got the impression that "Osgood knew what the students could take and what they couldn't take" and that "Foggy was clear about the applications."[24] Regarding infinitesimals and their acceptance by others on the Harvard faculty, however, Mac Lane later recalled that:

> ...Osgood was a world authority on functions of one and of several complex variables (a subject exemplifying

---

[22]See Osgood, 1907.
[23]See Osgood, 1922.
[24]Private communication with the author, 1996.

6. Osgood's Teaching

rigor). But Foggy knew well the intellectual limitations of the Harvard undergraduate, and wrote his texts on calculus accordingly. After his retirement we still used one of these texts for Math A. I recommended the appointment of one Leonidas Alaoglu, Ph.D. Chicago, whose merits were known to me. He came as a Benjamin Peirce instructor, and was of course set to teaching Math A. In that distant time, undergraduate women students did not attend classes in the Harvard Yard. So on one later October day Leon came to his class in Sever Hall,

> "Gentlemen, we now come to Chapter IV, differentials and infinitesimals. Take pages 138 to 184 between the thumb and fingers of the right hand. Tear them from the book!" He did so.[25]

Osgood is remembered by Mac Lane and others for this inclusion of infinitesimals in his calculus textbook. An examination of Osgood's conception of infinitesimals and why he felt they were a helpful pedagogical tool will shed some light on his approach to teaching.

What was an infinitesimal in Osgood's view? In his 1907 AMS presidential address, he emphasized that "the modern scientific conception is that an infinitesimal is a *variable* which it is generally useful to consider only for values numerically small and which, when the formulation of the problem has progressed to a certain state, is then allowed to approach zero as its limit." He contrasted this with the "archaic view of infinitesimals as small constants" which he dismissively called "the little zeros of former times."[26]. The mathematician Angus Taylor, who spent his faculty career at UCLA, served as Provost of the University of California system, and retired as Chancellor of U.C. Santa Cruz in 1977, had Osgood as a teacher during his undergraduate years at Harvard. He recalls

---
[25] See Mac Lane, 1996, pp. 1470–1471.
[26] See Osgood, 1907b, p. 451.

Osgood calling these archaic infinitesimals "those horrible little zeros."[27]

Osgood first made his views on infinitesimals clear in a 1903 paper entitled "The integral as the limit of a sum, and a theorem of Duhamel's."[28] He defined an infinitesimal as a variable whose limit is zero and he made clear that he was deeply concerned about rigor:[29]

> The treatment of the application of [the fundamental theorem of calculus and the corresponding theorem for multiple integrals] to problems in physics and mechanics, both in courses and in text-books on the calculus, is however far from satisfactory, the attitude of the mathematician often being that these applications are without value for mathematics, while for the physicist any reasoning is good enough which in the long run leads to the right formula. To throw rigor to the winds as soon as the hypnotic influence of the environment of a course in pure mathematics ceases is to regard rigor as a frill or as a luxury, as something having at most aesthetic reasons for existence,— not as a habit of thought, of practical value in research work. On the other hand the method to which this paper is devoted offers a valuable means of training students to appreciate the meaning of the integral calculus, and moreover a means which is alike valuable to students of applied and to students of pure mathematics. It quickens interest and develops power. The method appears to be due to Duhamel...

Osgood had a secondary reason for championing Duhamel's theorem—he wrote that "it was introduced into calculus teaching

---
[27] Private communication with the author, 1997.
[28] See Osgood, 1903c.
[29] See Osgood, 1903c, pp. 161–162.

6. Osgood's Teaching

in this country largely through Professor Byerly's, and Professor B.O. Peirce's text-books and teaching."[30] Both Byerly and Peirce had been Osgood's teachers at Harvard, and they were still his colleagues.

Duhamel's theorem is based on the idea that, given a sum of infinitesimals, in the limit any infinitesimal may be replaced by another infinitesimal that differs from the former by an infinitesimal of higher order. Osgood stated it this way:[31]

> Let $\alpha_1 + \alpha_2 + \cdots + \alpha_n$ be a sum of positive infinitesimals which approaches a limit when $n = \infty$. Let $\beta_1 + \beta_2 + \cdots + \beta_n$ be a second sum of positive infinitesimals which differ respectively from the infinitesimals of the first sum by infinitesimals of higher order; i.e. let
>
> $$\lim_{n=\infty} \frac{\beta_i}{\alpha_i} = 1.$$
>
> Then the second sum approaches a limit when $n = \infty$, and this limit is the same as that of the first sum:
>
> $$\lim_{n=\infty} (\alpha_1 + \alpha_2 + \cdots + \alpha_n) = \lim_{n=\infty} (\beta_1 + \beta_2 + \cdots + \beta_n).$$

Osgood gave an example to show how Duhamel's theorem is of use in applications to physics. He found the attraction of a straight rod with varying density on a particle in line with the rod. He divided the rod into segments of length Δx_i, and called the part of the attraction of the rod due to the ith segment ΔA_i. The total attraction was then the sum of the ΔA_i. He used the physical law that states that the attraction between two particles is given by the force:

$$f = \kappa \frac{mm'}{r^2}$$

---

[30] See Osgood, 1903c, p. 162.
[31] See Osgood, 1903c, p. 163.

where the m and m' are the masses of the objects, r is the distance between them, and κ is a gravitational constant. He argued that the actual force of attraction of the ith segment was greater than it would be if all the mass were concentrated at the far end of the ith segment and less than it would be if all the mass were concentrated at the near end of the ith segment. So if ρ' and ρ'' are the maximum and minimum densities on the ith segment:

$$\kappa \frac{m\rho_i' \Delta x_i}{x_i^2} < \Delta A_i < \kappa \frac{m\rho_i'' \Delta x_i}{x_{i-1}^2}.$$

All three terms above are infinitesimals and Osgood argued that the left and right hand infinitesimals differ from the middle one by an infinitesimal of higher order by dividing through by the left hand infinitesimal and examining the limit. This then showed that the condition, $\lim_{n=\infty} \frac{\beta_i}{\alpha_i} = 1$, of Duhamel's theorem was met. Applying it with the ΔA_i replaced by appropriate infinitesimals of higher order gave,

$$f(x_i) = \kappa \frac{m\rho_i \Delta x_i}{x_i^2}.$$

Osgood concluded that:

$$A = \lim_{n \to \infty} \sum_{i=1}^{n} \Delta A_i = \lim_{n \to \infty} \sum_{i=1}^{n} \kappa \frac{m\rho_i \Delta x_i}{x_i^2} = \int_a^b \kappa \frac{m\rho}{x^2} dx.$$

If, in particular, ρ is constant then the mass M of the rod is $\rho(b-a)$ and

$$A = \kappa m \rho (\frac{1}{a} - \frac{1}{b}) = \kappa \frac{mM}{ab}.$$

Osgood gave this and other examples to show how, in his view, infinitesimals could be useful in bringing rigor to applications in physics, without inflicting $\varepsilon - \delta$ proofs on students. But he went on to give a warning. In some problem situations, it was too easy to assume that a determined infinitesimal approached zero as its limit. He provided an example where an uncritical use of Duhamel's theorem led to a wrong answer. For $y = 1 + n^2 x e^{-n^2 x^2}$ as $n \to \infty$,

6. Osgood's Teaching

improper use of Duhamel's theorem to calculate the limit of the areas under the curve from 0 to 1 led to an answer of 1, the area under the limiting curve $y = 1$. This and other difficulties he described led Osgood to revise Duhamel's theorem:[32]

Let
$$\alpha_1 + \alpha_2 + \cdots + \alpha_n \qquad (A)$$
be a sum of infinitesimals and let α_i differ uniformly by an infinitesimal of higher order than Δx_i from the summand $f(x_i)\Delta x_i$ of the definite integral
$$\int_a^b f(x)dx \qquad (B)$$
of the function $f(x)$, this function being continuous throughout the interval $a \le x \le b$. Then the sum (A) approaches a limit when $n = \infty$, and the value of this limit is the definite integral (B):
$$\lim_{n=\infty} \sum_{i=1}^n \alpha_i = \int_a^b f(x)dx.$$

Osgood did not advocate use of this more complicated version of the theorem for all students. He believed that calculus students should use Duhamel's original theorem and that his own version should be used only in higher level mathematics courses or in mathematical physics courses when questions of uniform convergence arose naturally.[33]

In summary, in his beliefs about infinitesimals, Osgood was motivated by a desire for students to work with as much rigor as they could, by his view that infinitesimals arose naturally in applications to physics, and by his wish to stamp out use of "those horrid little

---
[32] See Osgood, 1903c, p. 173.
[33] See Osgood, 1903c. p. 178.

zeros" recalled by Angus Taylor. Taylor himself was influenced by Osgood's view of infinitesimals when he wrote both his *Calculus* with Sherwood[34] and his *Advanced Calculus*.[35] Taylor recalled:

> When I came to write a text-book on calculus, I decided to avoid entirely the word infinitesimal, but to retain the reason for introducing what I ultimately called Duhamel's principle, so that the limits of certain sums would be rigorously known to be certain definite integrals.... On p. 324 of the 3rd edition [of *Calculus*] I called the principle Osgood's form of Duhamel's principle.[36]

Taylor's books were widely used, but he also recalled that "not many other writers of calculus texts ventured to use Duhamel."[37] Osgood's teaching philosophy thus influenced his student Taylor, especially in relation to the use of infinitesimals to encourage a level of rigor appropriate to students. Taylor thereby transmitted Osgood's influence to the subsequent generations of students who used Taylor's own texts.

6.2 Lehrbuch der Funktionentheorie

Osgood's two volume treatise on the theory of functions, *Lehrbuch der Funktionentheorie*, was a text at an entirely different level. The first volume treated functions of one complex variable and the second treated functions of several complex variables. The inspiration for the volumes probably began when Felix Klein asked Osgood to write an article for the *Enzyklopädie der Mathematischen Wissenchaften*, in which Klein wanted to summarize mathematical

---
[34] See Sherwood and Taylor, 1942.
[35] See Taylor, 1955.
[36] Private communication with the author, 1996.
[37] Private communication with the author, 1997.

6. Osgood's Teaching

research up to 1900. Osgood's article "Allgemeine Theorie der analytischen Funktionen a) einer und b) mehrerer komplexen Grössen" (General Theory of Analytic Functions in one and several complex variables) appeared in 1901, and represented what J.L. Walsh called a "deep, scholarly, historical report on the fundamental processes and results of mathematical analysis."[38] There he explained that he wished to give a systematic development of the theory of functions using the infinitesimal calculus as a foundation and keeping close contact with geometry and mathematical physics. He noted that rigorous treatments of functions of real variables were not lacking, and intended for the *Funktionentheorie* to bring to the topics of functions of one and several complex variables the same standards of rigor, using the most modern methods of proof.[39]

Volume 1 of the *Funktionentheorie* first appeared in 1907.[40] The first five chapters of this volume are devoted to providing the reader with the rigorous foundations of real analysis and set theory necessary to study the remaining nine chapters. Chapters six through fourteen are divided into three parts: four chapters on foundations for the general theory of functions of one complex variable, including the topics of Riemann surfaces and analytic continuation; three chapters on applications of the theory to elementary, periodic and special functions; and two chapters on the logarithmic potential, conformal mapping and uniformization. This final section contains Osgood's proof of the Riemann mapping theorem. The first part of the second volume came out in 1924, followed by a second edition in 1929. This second edition of the first part of volume two was combined with a new second part in 1932 to complete the volume. The first part of volume 2 develops the foundations for the theory of functions of several complex variables over the course of three chapters. The five chapters of the second part deal primarily with the

---

[38] See Walsh, 1989, pp. 82–83.
[39] See Osgood, 1928, fifth edition, pp. iii–iv.
[40] The second edition of volume 1 was published in 1912, the third in 1920, the fourth in 1923 and the fifth in 1928.

theory of Abelian integrals. Osgood wrote both volumes entirely in German.

R.C. Archibald wrote that Osgood started the *Funktionentheorie* in 1901[41]; the final section appeared in 1932, and a number of revised editions appeared in the intervening years. This work, therefore, must have required a great deal of Osgood's attention for over thirty years, indeed for most of his career. That effort was not wasted. Archibald called the *Funktionentheorie* "one of America's greatest contributions to the development of mathematics."[42] He confirmed that when Osgood began the work, "there was no comprehensive treatment of the field, in which all the tools of ordinary modern analysis were rigorously used," and highlighted the fact that, in addition to organizing the material that already existed, Osgood filled many gaps in the literature, both small and large, without pointing them out as gaps or publishing his new contributions separately.[43]

Other mathematicians echoed Archibald's opinion of the *Funktionentheorie*. Edgar Lorch, who received his Ph.D. from Columbia in 1933 and was on the faculty there until 1977, made reference to the *Funktionentheorie* when he recalled his own studies at Columbia:

> T.S. Fiske was a kind, courteous, and distinguished person. ...However, he had long ago given up his research activities, and it was an open secret that if one was to learn function theory, one had to do it on one's own. I don't remember many ε's appearing on the board.... On the complex level, he made us read what he affectionately called "my little book" (*Functions of a Complex Variable*,

---
[41]See Archibald, 1938, p. 154.
[42]See Archibald, 1938, p. 155.
[43]See Archibald, 1938, p. 154.

97 pp., John S. Wiley, 1907), but it was clear that to learn the subject one had to read Konrad Knopp or Osgood.[44]

> Lorch's impression of the importance of Osgood's treatise is again confirmed by J.L. Walsh who said of the *Funktionentheorie*:

> It was more systematic and more rigorous than the French traités d'analyse, also far more rigorous than, say, Forsyth's theory of functions. It was a [monument] to the care, orderliness, rigor, and didactic skill of its author. When G. Pólya visited Harvard for the first time, I asked him whom he wanted most to meet. He replied "Osgood, the man from whom I learned function theory"— even though he knew Osgood only from his book. Osgood generously gives Bôcher part of the credit for the *Funktionentheorie*, for the two men discussed with each other many of the topics contained in it. The book became an absolutely standard work wherever higher mathematics was studied.[45]

Forsyth's book, Theory of Functions of a Complex Variable[46], was reviewed by Osgood for the Bulletin of the AMS[47]. Osgood criticized Forsyth's book for giving "little heed" to matters such as "accuracy and rigor in analysis."[48] He also cautioned that "the book is not one that can safely be put into the hands of the immature student for a first introduction to the study of the theory of functions."[49] These seem to be fatal criticisms—not enough rigor for the truly serious student and at the same time inappropriate for the beginner.

---

[44] See Lorch, 1989, p. 156.
[45] See Walsh, 1989, pp. 83–84.
[46] See Forsyth, 1893.
[47] See Osgood, 1895.
[48] See p. 142.
[49] See p. 154.

Angus Taylor recalls the influence that the *Funktionentheorie* had on his studies and textbooks in similar terms:

> All of these things [theorems about continuous functions] are dealt with with perfect clarity and rigor in Chapter 1 of [Osgood's] *Lehrbuch der Funktionentheorie*. That is where I learned the basics of the theory of continuous functions of a real variable, when I took Osgood's famous course, Math 13, Theory of Functions of a Complex Variable... He lectured every detail of the course (in English, of course), but I could read German and I bought the book (Vol. 1, 5th ed. 1928).
>
> When I came to write my *Advanced Calculus* I put all of the theory I had learned from Chapter 1 of the *Lehrbuch* into my book. But by that time I also knew about complete ordered fields and I thought the students ought to know what I knew.[50]

G.D. Birkhoff summed up the significance of Osgood's *Encyklopädie* article and *Funktionentheorie* in his 1938 survey of 50 years of American mathematics for the 50th anniversary of the AMS:

> His [*Encyklopädie* article] represents the first careful and systematic presentation of the Riemannian point of view, which is dominant today, as against the earlier Weierstrassian approach, based on the use of power series. Osgood's subsequent *Funktionentheorie*, has provided a large part of the present mathematical world with its fundamental training in this field, and remains today an invaluable adjunct to other books emphasizing more recent developments.[51]

---

[50] Private communication with the author, 1997.
[51] See Birkhoff, 1938, p. 293.

6. Osgood's Teaching

Osgood's *Funktionentheorie* stands as one of his greatest accomplishments. It provided a rigorous, modern treatment of one of the most important mathematical topics of the time. It transmitted his expertise as a world authority in the field to subsequent mathematicians both at home and abroad. It must be considered one of his most significant contributions as an educator to both the American and European mathematical communities.

6.2.1 Later textbooks

Two of Osgood's later textbooks grew out of his *Funktionentheorie*. He wrote *Functions of a Real Variable* and *Functions of a Complex Variable* based on lectures he gave in 1934 and 1935, following his retirement from Harvard, while serving as Professor of Mathematics at what was then called the National University of Peking. *Functions of a Real Variable* was addressed to students who had completed an advanced calculus course. In the preface Osgood referred to the "Three Theorems on Continuous Functions" (his capitalization) which he proved in Chapter 3, namely that a continuous function on a closed interval is bounded, that if the function changes sign it vanishes at a point in the interval, and that it attains its bounds. These are the same three theorems that earlier generations of students had found in Édouard Goursat's *Cours d'analyse mathématique*[52] called Theorems A, B and C, and that later generations found in Michael Spivak's *Calculus* referred to as "Three Hard Theorems."[53] Osgood reviewed the first volume of Goursat's text in 1903 and the second[54] in 1908. Although he pointed out some minor deficiencies, Osgood was effusive in his praise of the volumes, giving them high marks for both rigor and "good judgment in what [Goursat] expects of his readers," high praise indeed from one who constantly repeated these two themes.[55] It is therefore possible

---
[52] See Goursat, 1902.
[53] See Spivak, 1980, Chapter 7.
[54] See Goursat, 1905.
[55] See Osgood, 1908, p. 120.

that Osgood got the idea for the grouping of these three theorems from Goursat. Osgood later encouraged E.R. Hedrick to translate both volumes of Goursat into English. The result was *A Course in Mathematical Analysis*.[56]

Osgood also wrote *Functions of a Complex Variable* for the student who had completed a course in advanced calculus. However, he considered this book, not *Functions of a Real Variable*, the most appropriate textbook for a first course in higher analysis, calling it an "introduction to the most important methods and results of Modern Analysis."[57] To his colleagues who might have felt that real variables was the more elementary subject and therefore the suitable starting point for higher analysis, Osgood explained that he provided some needed material on real variables in the text but that:

> ...it must be borne in mind that the student is only just emerging from the Calculus, and it is reasonable to give him first what he can most readily receive. The theory of functions of real variables, if carried beyond its rudiments, soon loses contact, for the beginner, with the broader fields of analysis, geometry, and physics. It is these contacts, this broader knowledge, with which the beginner should become familiar before he specializes too closely in that great field, while for the student of Physics such specialization does not, at least at the present stage, come into consideration.

In the above passage, Osgood again demonstrated, nearly at the end of his career, two of the chief preoccupations that had shaped his teaching from the start—concern for teaching the student material that he was capable of handling at his level, and the belief that subjects should be taught with their practical applications.

---

[56]See Goursat and Hedrick, vol. 1 in 1904 and vol. 2 in 1917.
[57]See Osgood, 1948, p. iii.

6. Osgood's Teaching

Fittingly for one who believed in the benefits of applications to physics, Osgood's last textbook was entitled *Mechanics*. In the preface, Osgood acknowledged a debt to his teacher B.O. Peirce "who first blazed the trail in his course, Mathematics 4, given at Harvard in the middle of the eighties."[58]

Through his widely used textbooks, his advocacy of calculus reform, and his approach to teaching, Osgood played a significant role as a shaper of undergraduate mathematics education at Harvard and in the United States. At a more advanced level, he participated in the mathematical training of both undergraduate and graduate students indirectly yet meaningfully through his *Lehrbuch der Funktionentheorie*. Osgood, however, was less successful in having an immediately personal influence on graduate students and others in the classroom. As we shall see in the next section, he was not an inspiring classroom presence and did not leave a direct legacy of Ph.D. students to the American mathematical community.

---

[58] See Osgood, 1937, p. viii.

The Textbooks and Treatises of William Fogg Osgood[59]

1. *Introduction to Infinite Series.* Cambridge, 1897. (#10)

2. *Lehrbuch der Funktionentheorie.* Teubner, Leipzig, 1907. (#35)

3. *A First Course in the Differential and Integral Calculus.* New York, Macmillan, 1907. (#36)

4. *Plane and Solid Analytic Geometry* (with W.C. Graustein). New York, Macmillan 1921. (#58)

5. *Elementary Calculus.* New York, Macmillan, 1921. (#59)

6. *Introduction to Calculus.* New York, 1922. A revision of item 3 above. (#59)

7. *Advanced Calculus.* New York, Macmillan, 1925. (#64)

8. *A Short Table of Integrals.* A revision of B.O. Peirce's work of the same name. Boston, 1929. (#68)

9. *Functions of a Real Variable.* Peking, 1936 and New York, Hafner 1938. (#72)

10. *Functions of a Complex Variable.* Peking, 1936 and New York, Hafner, 1938. (#73)

11. *Mechanics.* New York, Macmillan, 1937. (#74)

---

[59]The numbers in parenthesis refer to the Osgood bibliography in Archibald, 1980, pp. 155–158, which includes information about multiple editions and reprints.

6.3 Bôcher's Introduction to Higher Algebra

Osgood wrote Harvard's calculus books, but it was Maxime Bôcher who provided the complement to those texts by writing the algebra textbook, even though algebra was not his field. Garrett Birkhoff called the textbooks of the two men "landmark texts," and noted that they were important in his own education:

> "Thus, I learned calculus from Osgood's books on the subject, and he was co-author of the book from which I learned analytic geometry (he also wrote one on mechanics). Bôcher was co-author with my high school teacher Harry Gaylord of a text on trigonometry, but it was his *Introduction to Higher Algebra* (1907), later translated into German and Russian, that was most famous. When Saunders Mac Lane and I were writing our *Survey of Modern Algebra*, I had this book and H.B. Fine's *College Algebra* much in mind, as the books whose substance we should reformulate axiomatically before emphasizing the general theories of (abstract) groups, rings and fields.[60]

Introduction to Higher Algebra appeared in 1907—the same year Osgood produced his first calculus text. It was perhaps one of the last of the major advanced undergraduate/graduate algebra books to be used in the United States before the advent of modern abstract algebra. Bôcher acknowledged the influence of Kronecker and Frobenius on the character of the book, and gave special thanks to Osgood for his suggestions and to a former student, E.P.R. Duval who helped him prepare the book for publication.[61]

Paul Halmos wrote about his memories and opinions of 26 different mathematics textbooks, one of which was Bôcher's *Higher Al-*

---
[60] See Birkhoff, 1977, p. 34.
[61] See Bôcher, 1907, p. vi.

gebra. He expressed no fondness for the book, and perhaps unfairly judged it from a modern viewpoint. But excerpts from his lengthy description give some idea of the book's character (the brackets are due to Halmos):

> From the point of view of 80 years after the first appearance of the book, the preface makes curious reading. It begins this way. "An American student approaching the higher parts of mathematics usually finds himself unfamiliar with most of the main facts of algebra, to say nothing of their proofs. [That sentence could have been written in 1987, couldn't it?] Thus he has only a rudimentary knowledge of systems of linear equations, and he knows next to nothing about the subject of quadratic forms. Students in this condition, if they receive any algebraic instruction at all, are usually plunged into the detailed study of some special branch of algebra, such as the theory of equations or the theory of invariants... [but that could surely not have been written in 1987]."
>
> The preface goes on to explain that a part of the purpose of the exercises at the ends of sections is to supply the reader with at least the outlines of important additional theories; as illustrations Bôcher mentions Sylvester's Law of Nullity, orthogonal transformations, and the theory of the invariants of the biquadratic binary form. Surely no modern author of a book on linear algebra would dare to relegate the first two of those to the exercises, and probably many modern authors of books on linear algebra have no idea of what the third one is all about....
>
> By "higher" algebra Bôcher meant a lot more than linear algebra; the lion's share of the book, the central chapters VII–XIX treat mainly invariant theory, bilinear and quadratic forms, and polynomials (including the theory of symmetric polynomials). Linear algebra comes back in a blaze of glory in the last three chapters...

Very few people still remember the book, and their memories of it are not always affectionate. May it rest in peace.[62]

Although Halmos did not like *Higher Algebra*, it enjoyed a great deal of success. It was reprinted at least 14 times, including a printing as late as 1937. Hans Beck translated *Higher Algebra* into German—Teubner published the translation in 1910. A.G. Kurosh translated the German into Russian and the Russian version was published in 1933.[63] Arthur Ranum, who reviewed *Higher Algebra* in 1910, found it to be "an ideal text-book, as I have found by actual trial in the class room." He also considered the fact that a German translation existed to be "a distinct compliment, not only to Professor Bôcher, but to American scholarship as well."[64]

---

[62] See Halmos, 1988, pp. 141–143.
[63] See Archibald, 1938, p.166.
[64] See Ranum, 1910, p. 523.

The Textbooks, Treatises and Monographs of Maxime Bôcher[65]

1. *Ueber die Reihenentwickelungen der Potentialtheorie.* Prizewinning doctoral dissertation at Göttingen, 1891. Elaborated and republished, Leipzig, 1894. (# 2, #13)

2. *Regular Points of Linear Differential Equations of the Second Order.* Part of AMS Colloquium Lectures. Cambridge, 1896. (#19)

3. Revision and enlargement of *The Elements of Plane Analytic Geometry* by George R. Briggs. New York, 1903. (#44)

4. *Introduction to Higher Algebra.* New York, Macmillan, 1907. (#56)

5. *An Introduction to the Study of Integral Equations.* Cambridge, England, 1909. (#59)

6. *Trigonometry with the Theory and Use of Logarithms* (with H.D. Gaylord). New York, 1915. (#75)

7. *Plane Analytic Geometry with Introductory Chapters on the Differential Calculus.* New York, 1915. (#76)

8. *Leçons sur les Méthodes de Sturm dans la Théorie des Équations Différentielles Linéaires et leurs Développements Modernes, professées à la Sorbonne en 1913–1914.* Collected and edited by G. Julia. Paris, 1917.

---

[65]The numbers in parenthesis refer to the Bôcher bibliography in Archibald, 1980, pp. 164–166, which includes information about multiple editions and reprints.

6.4 The Students of Osgood and Bôcher

Osgood fulfilled his duties as a responsible educator through his influential undergraduate textbooks and through his *Funktionentheorie,* which was seen as essential to advanced mathematical training. However, he supervised only four doctoral students in the 43 years he served at Harvard: C.W. McG. Black, L.D. Ames, E.H. Taylor, and G.R. Clements, the latter jointly with C.L. Bouton. In contrast, Bôcher supervised 17 in a much shorter career of 24 years, including J.W. Glover, M.B. Porter, F.H. Safford, D.F. Campbell, O. Dunkel, D.R. Curtiss, W.B. Ford, W.H. Roever, W.C. Brenke, F. Irwin, C.N. Moore, G.C. Evans, T. Fort, L.R. Ford, M.T. Hu, L. Brand, and C.N. Reynolds jointly with G.D. Birkhoff.[66]

Although Bôcher had many more doctoral students than Osgood, their joint contribution to the pool of mathematics Ph.D.s in the United States still pales in comparison to that of their counterparts at the University of Chicago. In the period 1896 through 1929, E.H. Moore supervised the doctoral dissertations of 29 students. Moore's first doctoral student, Leonard Eugene Dickson went on to supervise 64 doctoral dissertations at the University of Chicago from 1901 to 1937.[67] The University of Chicago clearly began producing the country's mathematics doctorates much more quickly than Harvard. Harvard, however, was at a disadvantage. It was an established school with traditions and faculty not yet well-oriented toward graduate education in 1900, while the University of Chicago had the advantage of having been structured to emphasize graduate education right from its start in 1892.

Osgood's doctoral students were undistinguished in their mathematical accomplishments, with the possible exception of E.H. Taylor. Taylor co-authored, with Osgood, the 1913 paper "Conformal

---

[66] See Archibald, 1938, pp. 153, 163.
[67] See Archibald, 1938, pp. 146, 185.

transformations on the boundaries of their regions of definition."[68] That important paper took Osgood's work on the Riemann mapping theorem a step further by examining behavior at the boundary.

Bôcher's doctoral students became more active members of the American mathematical community. The following are examples of how Bôcher's students were active, particularly in the work of the American Mathematical Society. J.W. Glover served as a council member of the AMS. M.B. Porter was also an elected member of the AMS council and acted as one of seven assistant editors of the AMS *Transactions* in 1902–1903. O. Dunkel assisted E.R. Hedrick with his translation/adaptation of the second volume of Goursat's *A Course in Mathematical Analysis*. D.R. Curtiss served on various AMS *Bulletin* committees and as its managing editor from 1937–1938. He chaired the Chicago Section of the AMS and was AMS Vice President at the national level in 1918. From 1914 to 1919 Curtiss held the title of editor of the AMS *Transactions*, and was managing editor for the last two of those years. He had a distinguished career at Northwestern University. W.B. Ford served as a member of the AMS council, as associate and assistant editor of the *Bulletin*, and as chair of the AMS Chicago Section. W.H. Roever held the chair of the AMS Southwestern section three times, gave an AMS invited address at the 1924 Ames, Iowa meeting on "Some phases of descriptive geometry." He also did war work under O. Veblen's direction at Aberdeen, Maryland in 1918. Like Roever, W.C. Brenke was chair of the Southwestern section. C.N. Moore gave AMS invited addresses in 1930 and 1933, respectively at Columbia, Missouri on "Types of series and types of summability" and at Chicago "On the Cesàro means in determining criteria for Fourier's constants." Moore acted as a cooperating editor of the *Transactions* from 1917 to 1935 and as AMS vice president in 1926–1927. Tomlinson Fort served the AMS as a *Bulletin* editor, member of the council and Associate Secretary (1931–1932). He enjoyed a fruitful career at Georgia Tech. L.R. Ford had a term

---

[68]See Osgood and Taylor, 1913.

6. Osgood's Teaching 173

as AMS vice president in 1946–1947 and taught at Rice Institute. His *Automorphic Functions* [69] remained for many years a standard reference.[70]

Bôcher's student G.C. Evans was at Rice University in Houston. Along with Veblen, Evans gave the 8th series of AMS Colloquium Lectures at Harvard in 1916 entitled "Functionals and their Applications, Selected Topics including Integral Equations." He served as editor of the *American Journal of Mathematics* from 1927 to 1935. He was AMS vice president in 1925–1926 and gave an AMS invited address (Southwestern section) in 1926 surveying "discontinuous boundary value problems for Laplace's equation in two-dimensional series." His 1936 AMS invited address treated "Methods of modern analysis in potential theory." He was president of the AMS in 1939–1940, and was a member of both the American Academy of Arts and Science and the National Academy of Sciences. Evans spent his years as a mathematician at Rice from 1912 to 1934 and at University of California, Berkeley from 1934 to 1954, where he led the department to high achievement.[71]

Clearly Osgood did not directly contribute to the growth of the American mathematical community by producing new doctorates in mathematics—Bôcher was the one to whom this task fell in the early years of the century. Some clues to at least one of the reasons for Osgood having had only a more indirect influence on the education of the next generation of new mathematicians can be found in the recollections of former students and colleagues.

The most prominent of Osgood's and Bôcher's former undergraduates got his doctorate at the University of Chicago with E.H. Moore. George David Birkhoff's first year at Chicago was as an undergraduate in 1902–1903. He recalled:

---

[69] See Ford, 1929.
[70] See Archibald, 1938, Birkhoff, 1989, p. 18, Parshall and Rowe, 1994, p. 437.
[71] See Archibald, 1938 and Pitcher, 1988.

> The year following I went to Harvard, with Moore's approval, for two years of study. There I learned more analysis, particularly from Osgood and Bôcher. I found Bôcher's lectures the equal of Bolza's in lucidity and superior in placing important points in high relief. It was only later, however, that I came to realize how much I owed to Bôcher for his suggestions, for his remarkable critical insight, and for his unfailing interest in the often crude mathematical ideas which I presented.[72]

G.D. Birkhoff's son Garrett also wrote about the importance of the influence, particularly of Bôcher, on his father:

> It was presumably under the stimulus of Bôcher (and perhaps Osgood) that [G.D. Birkhoff] wrote his first substantial paper (*Trans. Amer. Math. Soc.* 7 (1906), 107–36), entitled "General mean value and remainder theorems." The questions raised and partially answered in this are still the subject of active research. Moreover, his Ph.D. thesis, on expansion theorems generalizing Sturm-Liouville series, was also stimulated by Bôcher's ideas about such expansions, at least as much as by those of his thesis adviser, E.H. Moore, about integral equations.[73]

These recollections of G.D. Birkhoff and his son provide the hint that Osgood was not the most inspirational of teachers, particularly when compared to his colleague Bôcher. They also provide two examples of a characteristic pattern, referring to "Osgood and Bôcher" in the first quotation and "Bôcher (and perhaps Osgood)" in the second. The two men are frequently named together—they were often viewed as the Harvard team.

---
[72] See Birkhoff, 1938, pp. 274–275.
[73] See Birkhoff, 1989, p. 26.

6. Osgood's Teaching

G.D. Birkhoff returned to Harvard to join the faculty in 1912 as an already internationally acclaimed mathematician, for his proof of Poincaré's last geometric theorem among other work. (Bôcher had just given an invited address to the International Congress of Mathematicians in Cambridge, England, in which he devoted much time to Birkhoff's work on boundary value problems for ordinary differential equations.) Birkhoff and Osgood led an analysis seminar for research students that Birkhoff continued to lead until 1921.[74] From 1911 to 1937 Birkhoff supervised 37 doctoral students at Princeton, Yale, Harvard and Radcliffe.[75] Among them were J.L. Walsh, D.V. Widder, Bernard Koopman (Osgood's nephew), Marshall Stone, C.B. Morrey and Hassler Whitney. Walsh, Stone and Morrey all served as AMS presidents.[76] Walsh had asked Osgood to direct his thesis, "hopefully on some subject connected with the expansion of analytic functions, such as Borel's method of summation. [Osgood] threw up his hands, 'I know nothing about it.'."[77] So it was G.D. Birkhoff, not Osgood, who ultimately gave Walsh the graduate student his inspiration and support.

David V. Widder received his Ph.D. under Birkhoff in 1924. After six years on the faculty at Bryn Mawr, Widder returned to Harvard where he did research in analysis and wrote an influential advanced calculus text. In his nineties, he reminisced that:

> It is true that I have had the good fortune to be taught by or to have had contacts with many of the mathematical giants of the era, and do remember them with admiration and affection. For example, a portrait of Maxime Bôcher hangs now above my desk....
>
> The two courses that I remember most clearly that [1916 freshman] year were Analytic Geometry under Bôcher and Inorganic Chemistry under E.P. Kohler. It was a novel ex-

---
[74] See Birkhoff, 1989, pp. 16–27.
[75] See Archibald, 1938, p. 214.
[76] See Birkhoff, 1989, p. 27.
[77] See Walsh, 1989, p. 84.

Figure 6.2: G. D. Birkhoff

perience and somewhat exciting to be using a text that the professor had written: Bôcher's *Analytic Geometry*. Perhaps I did not appreciate at the time that a world famous mathematician had condescended to take a Freshman class. But I came to admire him and to become enamored with the subject.... In my Sophomore year I was lucky again. I had Modern Geometry under Bôcher. In the first weeks he had us discovering properties of the ellipse from familiar ones for the circle by use of affine transformations. This was just a foretaste of the marvels to come. I think it was the influence of this course by this instructor that determined for me the choice of a career. In any case I determined to take any course Bôcher offered in later years. The same year I studied Calculus under another famous mathematician, teaching from his own text.

6. Osgood's Teaching

> Professor W.F. Osgood had a less inspiring style. I recall that he gave us good advice, ignored by most, on how to prepare a paper. You were to fold it down the middle, put a first draft on the right, corrections on the left. He used rubber finger caps to hold chalk. On the whole I would describe him as somewhat imperious.[78]

Widder thus provided a piece of evidence to confirm the impression given by the Birkhoffs that Osgood was the less inspirational teacher of the two Harvard team members. Widder's remarks also provide an indicator that Osgood demonstrated an interest in helping his students acquire good study skills. This impression is confirmed by Osgood's undergraduate student, Angus Taylor.

Taylor named Osgood as the teacher who most influenced him at Harvard, but Bôcher was no longer there when Angus Taylor arrived to do his undergraduate studies. Taylor had weekly mathematics tutorial sessions with J.L. Coolidge but took all his formal mathematics courses with G.D. Birkhoff, E.V. Huntington, Marston Morse, Marshall Stone and Osgood, who was nearing retirement. Taylor remembers Osgood's writing and teaching as "systematic and clear." In his treatment of students, Osgood maintained "a certain formality" but was "attentive and kindly."[79] Like Widder, Taylor recalled that Osgood also enjoyed dispensing advice to students about studying mathematics, writing that:

> He was fond of uttering advice and opinions while teaching. He would say, "The clearer the teacher makes it, the worse it is for you. You must work things out for yourself and make the ideas your own.

Taylor added:

---
[78] See Widder, 1988, pp. 79–80.
[79] See Taylor, unpublished manuscript.

About studying something that seemed difficult his dictum was: Read it over a couple of times and sleep on it. Then, when you think you have grasped the situation, tell yourself the whole story in your own words, perhaps by writing it out on a scrap of paper while on the subway to Boston." In his books he liked to state problems in ways that required the student to enter into the process of formulating the problem sharply and clearly. He regarded this as an important part of the training of a mathematician. Some students, and teachers too, hated his books on that account.[80]

Thus, for students like Taylor, Osgood's classroom teaching could have some positive influence. If he was not dynamic and inspiring, he at least paid kind attention to the students and demonstrated concern for their learning process.

Norbert Wiener, who had gotten his Ph.D. at Harvard, benefited from Osgood's influence to a small extent. Wiener's father was a friend of Osgood and Wiener recalled being playmates with Osgood's sons. Recounting the months after he was discharged from military service at Aberdeen, Maryland in February 1919, Weiner wrote:

After several months of newspaper hack writing, I composed a couple of scientific articles on algebra which were good enough in their own way, but which have remained completely off the beaten track. Then Professor W.F. Osgood of Harvard secured me an appointment as instructor in the department of mathematics of the Massachusetts Institute of Technology.

---

[80]See Taylor, 1984, p. 607.

In characteristic fashion, a few paragraphs later, Wiener felt the need to modify the credit he had given Osgood for his assistance:

> I have perhaps been insufficiently grateful to Professor Osgood for the really good turn he did me in securing me the call to the Massachusetts Institute of Technology—or M.I.T., as it is more frequently known. There were, however, certain offsets to this act of kindness. For one thing, I never felt that I had earned any real esteem from him, nor did I feel that he had made me welcome at Harvard. Furthermore, jobs were plentiful with the resumption of normal life after the war.[81]

How, then, can Osgood's contribution to the mathematical community as an educator be summed up? As a Harvard educator, he clearly showed strength as a concerned and responsible teacher who demonstrated an abiding interest in undergraduate education. His approach to teaching was marked by a career-long preoccupation with appropriate standards of rigor and choices of material for the level of the student, and by a belief in the value of teaching practical applications, particularly from physics, along with theory. Osgood's undergraduate textbooks and participation in shaping the undergraduate curriculum, particularly calculus, brought a more modern approach to generations of students of mathematics at Harvard and elsewhere.

Osgood demonstrated less success in classroom teaching and in the guidance of graduate students. He was not an inspiring classroom presence for many students, and he played no significant role in directing the dissertations of doctoral students. Luckily Harvard could count on Maxime Bôcher, the other member of the Harvard team, admirably to fulfill the role of inspiring classroom teacher, and on Bôcher and later G.D. Birkhoff to produce the Ph.D.s.

---

[81] See Wiener, 1956, pp. 30–31.

Regardless of his shortcomings as an educator at Harvard, however, Osgood influenced a number of American and even European mathematicians through his treatise *Funktionentheorie*. He participated with E.H. Moore and others in the dialogue on the form that undergraduate mathematics education should take;[82] he was a member of the AMS committee that recommended college entrance requirements;[83] and served as one of the American commissioners of the International Commission on the Teaching of Mathematics. These national and international activities extended Osgood's contribution to establishing educational traditions and standards beyond Harvard.

Finally, Osgood's function as a mathematical role model for students must not be discounted—as a world-class researcher, he raised the standard for American mathematicians. As noted earlier, Coolidge summed up Osgood and Bôcher as teachers by concluding that, "the principal contribution of these two men was to establish traditions and standards as a guide to the future."[84] Both Osgood and Bôcher established these traditions and standards through their textbooks, their approaches to teaching, and their internationally renowned research.

---

[82]See Maltbie, 1900.
[83]See their recommended requirements in the *Bulletin* of the AMS, 10: pp. 74–77, 1903.
[84]See Coolidge, 1930, p. 251.

The Educational Writings of William Fogg Osgood[85]

1. The undergraduate mathematical curriculum. (Report of a discussion at the AMS summer meeting.) *Bulletin of the American Mathematical Society*, 7:18–21, 1900. (#20)

2. The calculus in our colleges and technical schools. *Bulletin of the American Mathematical Society*, 13: 449–467, 1907. (#34)

3. *Report of the American Commissioners of the International Commission on the Teaching of Mathematics* (with D.E. Smith and J.W.A. Young). ICT Math., *Amer. Report.* Washington, 1912 and U.S. Bureau of Education, *Bulletin*, 1912. (They wrote other reports in this series as well.) (#40)

4. Suggestions and advice to examiners in mathematics. College Entrance Examination Board, *Annual Report*, pp. 52–55, 1920. (#57)

5. *Report... upon Elementary Algebra, Advanced Algebra, and Plane Trigonometry,* by Osgood as Chairman of a Commission on College Entrance Requirements, to the College Entrance Examination Board. New York, 1922 and 1923. (#62)

6. *Report... on the Requirements in Geometry,* by Osgood as Chairman of a Commission on College Entrance Requirements, to the College Entrance Examination Board. New York, 1922 and 1923. (#62)

7. *On the New Type of Examinations in Elementary Algebra prepared at the Request of the C.E.E.B. by the Institute of Educational Research, Teachers College, Columbia U.*, 1922. (#62)

---

[85] The numbers in parenthesis refer to the Osgood bibliography in Archibald, 1980, pp. 155–158.

The Educational Writings of Maxime Bôcher[86]

1. The fundamental conceptions and methods of mathematics. *Bulletin of the American Mathematical Society*, 11:115–135, 1904. Also in *Congress of Arts and Science, Universal Exposition, St. Louis, 1904*, ed. H.J. Rogers. Boston, 1: 456–473, 1905. (#48)

2. Graduate work in mathematics in universities and in other institutions of like grade in the United States. General report. U.S. Bureau of Education, *Bulletin* 6: 7–20, 1911. Also in *Bulletin of the American Mathematical Society*, 18:122–137, 1911. (#63)

3. Doctorates conferred by American universities. Letter to the editor, *Science*, 38: 546, 1913. (#67)

4. Mathématiques et mathématiciens français. *Rev. Intern. de l'Enseignement*, 67:20–31, 1914. (#71)

5. *Syllabus of a Brief Course in Solid Analytic Geometry*, Lancaster, 1910. (#78)

---

[86]The numbers in parenthesis refer to the Bôcher bibliography in Archibald, 1980, pp. 164–166.

6. Osgood's Teaching

Osgood and Bôcher came into their professional maturity at a very special time in the development of mathematics in the United States. Although this book has focused a great deal on Osgood, they both became dynamic agents of the transformation of the American mathematical arena. It would never again be expected that an American mathematician at a major university would spend a career teaching elementary mathematics, and maybe a science, to a small group of undergraduates. The modern American mathematician would instead be seen as a participant in a larger mathematical community, would be a responsible educator of the next generation of mathematicians and other students, and would strive to do important mathematical research.

Osgood and Bôcher formed the nucleus of the Harvard team in shaping the transformation. They provided leadership and direction to the newly-emerged *community of mathematicians*; and they helped define the new *responsible educator*, both benefiting from and setting new traditions and standards at Harvard and nationally. They brought significant European attention to the new community in their roles as *top mathematical researchers*—attention which was a factor in legitimizing the mathematical community as a mainstream *research* community.

Bibliography

Ahlfors, Lars V.: Development of the Theory of Conformal Mapping and Riemann Surfaces Through a Century. In: *Contributions to the Theory of Riemann Surfaces, Centennial Celebration of Riemann's Dissertation*, Annals of Mathematics Studies No. 30, eds. L. Ahlfors, E. Calabi, M. Morse, L. Sario and D. Spencer, pp. 3-13. Princeton, New Jersey, Princeton University Press, 1953.

Ahlfors, Lars V.: *Complex Analysis, An Introduction to the Theory of Analytic Functions of One Complex Variable.* New York, McGraw-Hill, third edition, 1979.

Archibald, Raymond Clare: Benjamin Peirce: 1809-1880, Biographical Sketch and Bibliography, Oberlin, Ohio, MAA, 1925. In: *Benjamin Peirce: Father of Pure Mathematics in America*, ed. I. Bernard Cohen. New York, Arno Press, 1980.

Archibald, Raymond Clare: *A Semicentennial History of the American Mathematical Society.* New York, American Mathematical Society, 1938. Reprinted New York, Arno Press, 1980.

Arzelà, Cesare: Sulla integrabilità di una serie di funzioni, *Rendiconti, Accademia dei Lincei,* 1: 321-326, 1885.

Birkhoff, G.D.: The Scientific Work of Maxime Bôcher. *Bulletin of the American Mathematical Society*, 25: 197-215, 1919. (Reprinted in *A Century of Mathematics in America – Part II*, ed. Peter Duren, pp. 59-78. Providence, Rhode Island, American Mathematical Society, 1989.)

Birkhoff, G.D.: Fifty years of American mathematics. In: *Semicentennial Addresses of the American Mathematical Society*. New York, American Mathematical Society, 1938.

Birkhoff, Garrett and Mac Lane, Saunders: *Survey of Modern Algebra*. New York, Macmillan, 1941

Birkhoff, Garrett: Some Leaders in American Mathematics: 1891-1941. In: *The Bicentennial Tribute to American Mathematics 1776-1976*, ed. Dalton Tarwater, Mathematical Association of America, 1977.

Birkhoff, Garrett: Mathematics at Harvard, 1836-1944. In: *A Century of Mathematics in America – Part II*, ed. Peter Duren, pp. 3-58. Providence, Rhode Island, American Mathematical Society, 1989.

Bliss, Gilbert Ames: Integrals of Lebesgue. *Bulletin of the American Mathematical Society*, 24: 1-47, 1917.

Block, N. Henry: Certain ancient physical apparatus belonging to Harvard College. *Harvard Alumni Bulletin*, 25, 1933.

Bôcher, Maxime: *Introduction to Higher Algebra*. New York, Macmillan, 1907.

Borel, Émile: *Leçons sur les Fonctions de Variables Réelles*. Paris, Gauthier-Villars, 1905.

Bottazzini, Umberto: *The Higher Calculus: A History of Real and Complex Analysis from Euler to Weierstrass.* New York, Springer-Verlag, 1986.

Bourbaki, Nicolas: *Éléments d'histoire des mathématiques.* Paris, Hermann, 1960.

Browder, Andrew: *Introduction to Function Algebras.* New York, W.A. Benjamin, 1969.

Bruce, Robert V.: *The Launching of Modern American Science, 1846-1876.* Ithaca, New York, Cornell University Press, 1987.

Byerly, William E.: Reminiscences (of Benjamin Peirce). In: Archibald, Raymond Clare: Benjamin Peirce: 1809-1880, Biographical Sketch and Bibliography, Oberlin, Ohio, MAA, 1925. In: *Benjamin Peirce: Father of Pure Mathematics in America*, ed. I. Bernard Cohen. New York, Arno Press, 1980.

Cajori, Florian: *The Teaching and History of Mathematics in the United States.* Washington, DC, Government Printing Office, 1890.

Cajori, Florian: *A History of Mathematics*, second edition. New York, Macmillan, 1924.

Cohen, Bernard I., ed.: *Benjamin Peirce: Father of Pure Mathematics in America.* New York, Arno Press, 1980.

Coolidge, Julian Lowell: Mathematics. In: *Development of Harvard University, 1869-1929*, ed. Samuel Eliot Morison, pp. 248-257. Cambridge, Massachusetts, Harvard University Press, 1930.

Cremin, Lawrence A.: *The Transformation of the School: Progressivism in American Education, 1876-1957.* New York, Alfred A. Knopf, 1961.

Cremin, Lawrence A.: *The Metropolitan Experience 1876-1980.* New York, Harper and Row, 1988.

Denjoy, Arnaud: Notice sur la vie et l'oeuvre de Henri Lebesgue. *Notices et Discours de l'Académie des Sciences,* 2, 1949.

Dirichlet, P.G.L.: *Vorlesungen über die im umgekehrten Verhältniss des Quadrats der Entfernung wirkenden Kräfte,* second edition, ed. F. Grube. Leipzig, 1887.

Eliot, Charles William: Inaugural Address as President of Harvard. 1869a. In: *Development of Harvard University, 1869-1929,* ed. Samuel Eliot Morison, pp. lix - lxxviii. Cambridge, Massachusetts, Harvard University Press, 1930.

Eliot, Charles William: The New Education: Its Organization. *Atlantic Monthly,* 23, 1869b.

Fiske, T.S.: Frank Nelson Cole. *Bulletin of the American Mathematical Society,* 23: 773-777, 1927.

Ford, Lester B.: *Automorphic Functions.* New York, McGraw-Hill, 1929.

Forsyth, A.R.: *Theory of Functions of a Complex Variable.* Cambridge, University Press, 1893.

Gray, Jeremy: *Linear Differential Equations and Group Theory from Riemann to Poincaré.* Boston, Birkhäuser, 1986.

Gibson, George A.: *An Elementary Treatise on the Calculus.* London, Macmillan, 1901.

Goursat, Édouard: *Cours d'analyse mathématique.* Vol. 1. Paris, Gauthier-Villars, 1902.

Goursat, Édouard and Hedrick, Earle Raymond: *A Course in Mathematical Analysis.* Vol. 1. Boston, Ginn and Company, 1904.

Goursat, Édouard: *Cours d'analyse mathématique.* Vol. 2. Paris, Gauthier-Villars, 1905.

Goursat, Édouard and Hedrick, Earle Raymond: *A Course in Mathematical Analysis.* Vol. 2. Boston, Ginn and Company, 1917.

Halmos, Paul. R.: Some Books of Auld Lang Syne. In: *A Century of Mathematics in America – Part I*, ed. Peter Duren, pp. 131-174. Providence, Rhode Island, American Mathematical Society, 1988.

Harkness, James: Review of *Introduction to Infinite Series,* by William F. Osgood. *Bulletin of the American Mathematical Society,* 4: 277-278, 1898.

Harnack, Axel: Ueber den Inhalt von Punktmengen. *Mathematische Annalen,* 25: 241-250, 1885.

Harnack, Axel: *Die Grundlagen der Theorie des logarithmischen Potentiales und der eindeutigen Potentialfunktion in der Ebene.* Leipzig, B.G. Teubner, 1887.

Haskins, Charles N.: Osgood's Calculus. *Bulletin of the American Mathematical Society,* 15: 457-462, 1909.

Hawkins, Thomas: *Lebesgue's Theory of Integration: Its Origins and Development*. Madison, The University of Wisconsin Press, 1970.

Hille, Einar: *Analytic Function Theory*. Vol. II. New York, Ginn and Company, second edition, 1973.

Hubbell, J.G. and Smith, R.W.: Neptune in America – Negotiating a Discovery, *Journal for the History of Astronomy*, 23(4): 261-291, 1992.

Hurwitz, W.A.: *Verhandlungen des ersten internationalen Mathematiker-Kongresses in Zürich vom 9, bis 11. August 1897*. Leipzig, 1898.

Jordan, Camille: Remarques sur les intégrales défines. *Journal de Mathématiques pures et appliquées,* 8: 69-99, 1892.

Jordan, Camille: *Cours d'analyse de l'École polytechnique*. Paris, Gauthier-Villars, second edition, 3 vols., 1893-1896.

Katz, Victor J.: *A History of Mathematics: An Introduction*. New York, HarperCollins College Publishing, 1993.

Klein, Felix: *Vorlesungen über die Entwicklung der Mathematik in 19. Jahrhundert*, 2 vols. Berlin, 1926-1927. Reprint (2 vols. in 1). New York, Chelsea Publishing Company, 1967.

Kline, Morris: *Mathematical Thought from Ancient to Modern Times*. New York, Oxford University Press, 1972.

LaDuke, Jeanne: The Study of Linear Associative Algebras in the United States, 1870-1927. In: *Emmy Noether in Bryn Mawr, Proceedings of a Symposium sponsored by the AWM in Honor of*

Emmy Noether's 100th Birthday, eds. Bhama Srinivasan and Judith Sally, pp. 147-159. New York, Springer Verlag, 1983.

Lang, Serge: *Complex Analysis*. New York, Springer-Verlag, second edition 1985.

Lebesgue, Henri: Intégrale, Longueur, Aire. *Thèses présentées a la Faculté des Sciences de Paris...* Milan, Imprimerie Bernardoni de C. Rebeschini & C., 1902.

Lebesgue, Henri: Sur la méthode de M. Goursat pour la résolution de l'équation de Fredholm. *Bulletin de la Société Mathématique de France*, 36: 3-19, 1908.

Lorch, Edgar R.: Mathematics at Columbia During Adolescence. In: *A Century of Mathematics in America – Part III*, ed. Peter Duren, pp. 149-161. Providence, Rhode Island, American Mathematical Society, 1989.

Lukes, Jaroslav; Malý, Jan; and Zajícek, Ludek:*Fine Topology Methods in Real Analysis and Potential Theory*, Lecture Notes in Mathematics No. 1189. New York, Springer-Verlag, 1986.

Mac Lane, Saunders: Comments on the Harvard Consortium Calculus Text, Letter to the Editor. *Notices of the American Mathematical Society*, 43: 1469-1471, 1996.

Mac Lane, Saunders: *Saunders Mac Lane, A Mathematical Autobiography*, Wellesley, Massachusetts, A K Peters, 2005.

Maltbie, W.H.: The undergraduate mathematical curriculum – Report of the discussion at the seventh summer meeting of the American Mathematical Society. *Bulletin of the American Mathematical Society,* 7: 14-24, 1900.

Mandelbrot, Benoit B.: *The Fractal Geometry of Nature*, New York, W.H. Freeman and Company, 1977, revised and augmented 1983.

May, Kenneth: Biographical essay. In: *Measure and the Integral by Henri Lebesgue,* edited and with a biographical essay by Kenneth O. May. San Francisco, Holden-Day, 1966.

Monna, A.F.: *Dirichlet's Principle: A Mathematical Comedy of Errors and Its Influences on the Development of Analysis.* Utrecht, Oosthoek, Scheltema & Holkema, 1975.

Montel, Paul: The Role of Families of Functions in Mathematical Analysis. In: *Great Currents of Mathematical Thought*, vol. 1 Mathematics: Concepts and Development, pp. 174-180. New York, Dover, second edition, 1971.

Moore, E.H.: On the foundations of mathematics. Presidential address before the American Mathematical Society, December 29, 1902. *Bulletin of the American Mathematical Society,* 9: 402-424, 1903.

Nehari, Zeev: *Conformal Mapping.* McGraw-Hill, 1952. Reprint. New York, Dover, 1975.

Newcomb, Simon: *Royal Society of Edinborough, Proceedings*, 22, 1880-1882.

Osgood, William Fogg: The Theory of Functions (Review of A.R. Forsyth's *Theory of Functions of a Complex Variable). Bulletin of the American Mathematical Society,* 1: 142-154, 1895.

Osgood, William Fogg: A Geometrical Method for the Treatment of Uniform Convergence and Certain Double Limits. *Bulletin of the American Mathematical Society,* 3: 59-86, 1896.

Osgood, William Fogg: Non-uniform Convergence and the Integration of Series Term by Term. *American Journal of Mathematics*, 19: 155-189, 1897a.

Osgood, William Fogg: *Introduction to Infinite Series.* Cambridge, Harvard, 1897b.

Osgood, William Fogg: Selected Topics in the General Theory of Functions. *Bulletin of the American Mathematical Society*, 5, 1898.

Osgood, William Fogg: On the Existence of the Green's Function for the Most General Simply Connected Plane Region. *Transactions of the American Mathematical Society*, 1: 310-314, 1900.

Osgood, William Fogg: Notes and errata for "On the Existence of the Green's Function for the Most General Simply Connected Plane Region." *Transactions of the American Mathematical Society*, 2: 484-485, 1901.

Osgood, William Fogg: A Modern English Calculus. Review of *An Elementary Treatise on the Calculus* by George A. Gibson. *Bulletin of the American Mathematical Society*, 8: 248-257, 1902.

Osgood, William Fogg: A Jordan Curve of Positive Area. *Transactions of the American Mathematical Society*, 4: 107-112, 1903a.

Osgood, William Fogg: On the transformation of the boundary in the case of conformal mapping. *Bulletin of the American Mathematical Society,* 9: 233-235, 1903b.

Osgood, William Fogg: The integral as the limit of a sum, and a theorem of Duhamel's. *Annals of Mathematics,* 4: 161-178, 1903c.

Osgood, William Fogg: A modern French calculus. *Bulletin of the American Mathematical Society,* 9: 547-555, 1903d.

Osgood, William Fogg: *Lehrbuch der Funktionentheorie.* First published, Leipzig, 1907a.

Osgood, William Fogg: The calculus in our colleges and technical schools (AMS Presidential Address). *Bulletin of the American Mathematical Society,* 13: 449-467, 1907b.

Osgood, William Fogg: *A First Course in the Differential and Integral Calculus.* New York, Macmillan, 1907c.

Osgood William Fogg: Goursat's *Cours d'analyse. Bulletin of the American Mathematical Society,* 15: 120-126, 1908.

Osgood, William Fogg and Taylor, E.H.: Conformal transformations on the boundaries of their regions of definition. *Transactions of the American Mathematical Society,* 14: 277-298, 1913.

Osgood, William Fogg: The Life and Services of Maxime Bôcher. *Bulletin of the American Mathematical Society,* 25: 337-350, 1919.

Osgood, William Fogg: *Introduction to Calculus.* New York, Macmillan, 1922.

Osgood, William Fogg: *Lehrbuch der Funktionentheorie*, Volume 1. Leipzig, Teubner, fifth edition, 1928.

Osgood, William Fogg: *Lehrbuch der Funktionentheorie*, Volume 2. Leipzig, Teubner, second edition, 1929.

Osgood, William Fogg: *Mechanics.* New York, MacMillan, 1937.

Osgood, William Fogg: *Functions of Real Variables.* New York, Hafner, 1938.

Osgood, William Fogg: *Functions of a Complex Variable.* New York, Hafner, 1948.

Parshall, Karen Hunger and Rowe, David E.: *The Emergence of the American Mathematical Research Community, 1876-1900: J.J. Sylvester, Felix Klein, and E.H. Moore.* Providence, Rhode Island, American Mathematical Society, 1994.

Parshall, Karen Hunger: *James Joseph Sylvester, Jewish Mathematician in a Victorian World.* Baltimore, The Johns Hopkins University Press, 2006.

Peano, Giuseppe: Sulla integrabilità della funzioni, *Atti della R. Accademia delle Scienze di Torino*, 18: 439-446, 1883.

Peano, Giuseppe: *Applicazioni geometriche del calcolo infinitesimale,* Bocca, Turin, 1887.

Peirce, Benjamin: *A System of Analytic Mechanics.* Boston, Little, Brown & Co., 1855.

Peirce, Benjamin: Linear Associative Algebra: New Edition with Addenda and Notes, by C.S. Peirce, Son of the Author, New York, 1882. In: *Benjamin Peirce: Father of Pure Mathematics in America,* ed. I. Bernard Cohen. New York, Arno Press, 1980.

Peirce, Benjamin Osgood: *Theory of the Newtonian Potential Function.* Boston, Ginn & Company, 1888.

Peirce, Benjamin Osgood: *A Short Table of Integrals.* Boston, 1929.

Peterson, Sven R.: Benjamin Peirce, Mathematician and Philosopher (Reprinted from *Journal of the History of Ideas,* vol. 16), New York, 1955. In: *Benjamin Peirce: Father of Pure Mathe-*

matics in America, ed. I. Bernard Cohen. New York, Arno Press, 1980.

Picard, Émile: *Les sciences mathématiques en France depuis un demi-siècle*, Paris, 1917.

Pitcher, Everett: *A History of the Second Fifty Years: American Mathematical Society, 1939-1988*. Providence, Rhode Island, American Mathematical Society, 1988.

Poincaré, Henri: Sur un théorème de la théorie générale des fonctions. *Bulletin de la Société Mathématique de France*, 11:112-125, 1883.

Poincaré, Henri: Sur l'uniformisation des fonctions analytiques. *Acta Mathematica*, 31: 1-63, 1908.

Poincaré, Henri: *Oeuvres de Henri Poincaré*. Vol. XI – Mémoires divers – hommages a Henri Poincaré. Paris, Gauthier-Villars, 1956.

Hubbell, John G. and Smith, Robert W. "Neptune in America: Negotiating a Discovery," *Journal for the History of Astronomy*, 1992.

Ranum, Arthur: Bôcher's *Higher Algebra*. *Bulletin of the American Mathematical Society*, 16: 521-523, 1910.

Reingold, Nathan: *Science in Nineteenth-Century America, A Documentary History*. New York, Hill and Wang, 1964.

Reingold, Nathan: Nathaniel Bowditch. In: *Dictionary of Scientific Biography*, ed. Charles Coulston Gillispie, pp. 368-369. New York, Charles Scribner's Sons, 1970.

Riemann, Bernhard: *Mathematische Werke.* Leipzig, B.G. Teubner, 1876.

Riesz, Frigyes and Sz.-Nagy, Béla: *Functional Analysis.* New York, Frederick Ungar Publishing Co., 1955. Dover edition 1990.

Rosenstein, George M.: American Calculus Textbooks of the Nineteenth Century. In: *A Century of Mathematics in America – Part III*, ed. Peter Duren, pp. 77-109. Providence, Rhode Island, American Mathematical Society, 1989.

Schlereth, Thomas J.: *Victorian America: Transformations in Everyday Life 1876-1915.* New York, HarperCollins, 1991.

Schoenflies, Arthur: Die Entwickelung der Lehre von den Punktmannigfaltigkeiten.*Jahresbericht der Deutschen Mathematiker-Vereinigung*, 8, 1899.

Sherwood, G.E.F. and Taylor, Angus: *Calculus.* New York, Prentice Hall, 1942.

Smith, David Eugene and Ginsburg, Jekuthiel: *A History of Mathematics in America Before 1900.* Chicago, Mathematical Association of America (with Open Court Publishing) Chicago, 1934.

Smith, H.J.S.: On the Integration of Discontinuous Functions. *Proceedings of the London Mathematical Society*, 6: 140-153, 1875.

Spivak, Michael: *Calculus.* Houston, Publish or Perish, second edition, 1980.

Susman, Warren: *Culture as History: The Transformation of American Society in the Twentieth Century.* New York, Pantheon, 1984.

Sutherland, W.A.: *Introduction to Metric and Topological Spaces.* New York, Oxford University Press, 1975.

Taylor, Angus: *Advanced Calculus.* Boston, Ginn and Co., 1955.

Taylor, Angus: A Life in Mathematics Remembered. *The American Mathematical Monthly,* 91: 605-618, 1984.

Thomas, M. Carey: The Bryn Mawr Woman. In: *The Educated Woman in America: Selected Writings of Catherine Beecher, Margaret Fuller, and M. Carey Thomas,* ed. Barbara M. Cross. New York, Teachers College Press, 1965.

Van Vleck, Edward Burr: Haskins's momental theorem and its connection with Stieltjes's problem of moments. *Transactions of the American Mathematical Society,* 18: 326-330, 1917.

Veysey, Laurence R.: *The Emergence of the American University.* Chicago, University of Chicago Press, 1965.

Walsh, Joseph, L: William Fogg Osgood. In: *A Century of Mathematics in America – Part II,* ed. Peter Duren, pp. 79-85. Providence, Rhode Island, American Mathematical Society, 1989.

Weierstrass, Karl: Über das sogenannte Dirichlet'sche Princip. Read July 1870. In: *Mathematische Werke,* 7 vols., vol. 2: 49-54. Berlin, 1854-1927. Reprint. Hildesheim, Georg Ohms Verlagsbuchhandlung and New York, Johnson Reprint Corporation, 1967.

Webber, Samuel: *Mathematics, Compiled from the Best Authors and Intended to Be a Text-book of the Course of Private Lectures on These Sciences in the University at Cambridge.* Boston, Harvard College, 1801.

Wermer, John: Polynomial Approximation on an Arc in $\mathbb{C}^3$, *Annals of Mathematics*, 62, 1955.

Wermer, John: *Banach Algebras and Several Complex Variables*, Chicago, Markham Publishing, 1971.

Widder, David V.: Some Mathematical Reminiscences. In: *A Century of Mathematics in America – Part I*, ed. Peter Duren, pp. 79-83. Providence, Rhode Island, American Mathematical Society, 1988.

Wiener, Norbert: *Ex-Prodigy, My Childhood and Youth*. Cambridge, Mass., The M.I.T. Press, 1964.

Wiener, Norbert: *I Am a Mathematician: The Later Life of a Prodigy*. New York, Doubleday, 1956.

Appendix A

Letters of Angus E. Taylor to Diann Porter

Angus E. Taylor earned his undergraduate degree from Harvard in 1933. He spent much of his career as a mathematician at UCLA, beginning in 1938, and including a long stint as chair of the mathematics department. In 1965, he took the newly created position of Vice President for Academic Affairs of the University of California system. He later served as Provost of the UC system, its highest academic officer, and then led UC Santa Cruz as Chancellor until his retirement in 1977. He chaired the University of California's Academic Council of the Academic Senate during the Free Speech movement of 1964–65 and is remembered as a peace-maker.

Professor Taylor asked that I acknowledge that he gave me the information in this appendix, some of which is also included in the body of the book, and let my readers know that he had great respect for Osgood as a teacher and mathematical scholar.

Professor Taylor passed away in 1999 at the age of 87, two years after this correspondence took place. The correspondence was both a personal pleasure for me, and a part of the research for my doctoral dissertation. I was and remain grateful to him for sharing his time and experience so generously.

Some of his reminiscences can be found in his article "A Life in Mathematics Remembered," *The American Mathematical Monthly*, 91: 605–618, 1984.

The following sections are transcripts of handwritten letters from Angus E. Taylor to Diann Porter.

A.1   January 25, 1997

Dear Ms Porter,

Your letter intrigues me because I do know quite a bit about Osgood and his work. But I am going to be quite tied down for at least a week, so this is certainly not my full answer to your letter.

I presume that Karen Parshall learned about my Harvard connection from the article I wrote for the Amer. Math Monthly vol. 91 (1984), 605–618.

I presume that you have consulted the article about Osgood in the 50th anniversary volumes of the AMS. I know all of his books, including the Lehrbuch der Funktionentheorie and a book about analysis published in China.

You may not know that he went to Reno to divorce his first wife, a German lady, and then married the woman who had been Mrs. Marston Morse.

Osgood was a powerful analyst, who came close to discovering the phenomena in point set theory that made Baire famous.

He was very nice to me—took me to dinner at the Harvard Club in Boston, commented favorably on my first book—Calculus (by Sherwood and Taylor).

I'll leave further information to a later letter, perhaps 2 or 3 weeks from now. But look at his paper about the limit of an infinite series of analytic functions that need not converge uniformly.

Yours faithfully,

/s/Angus E. Taylor

A.2 January 25, 1997

Dear Ms. Porter

This is a P.S. which I'm sending because the ideas in it came after I sealed up my first letter.

Garrett Birkhoff, the son of the great G.D.B. is listed in the directory of the AMS and MAA, but I've heard that he is not well. He was one year ahead of me at Harvard. You might try writing him. He would have known a lot about Osgood.

Your might also try Mrs. Joseph L. Walsh, the 2nd wife of Prof J.L. Walsh, whom I knew at Harvard.[1]

She outlived Joe. I met her and know her quite well through my brother O.H. Taylor, a prof. of Econ at Harvard who retired and

---

[1] *Ed.* Address omitted.

taught at Vanderbilt Univ. for a few years. I knew her as Liz. We used to exchange Christmas cards, but that tapered off. I am 85. She was younger.

Finally, I presume you know the Bi-Centennial Tribute to American Mathematics 1 1776–1976. See G. Birkhoff's article, especially pp. 32–34.

I've tried to pick out, from the publication list of Osgood's work (in vol 1 of the semicentennial pub.), the remarkable paper in which Osgood showed that if a sequence of analytic functions converges pointwise, but not necessarily uniformly, that in any neighborhood of a point of convergence there is an open set in which the limit function is holomorphic. I have notes about that somewhere, but not readily at hand. Possibly item 23 in the bibliography.

Au revoir,

/s/AET.

A.3 February 15 and 16, 1997

Dear Diann Porter,

I started to write this letter on 2/15, but am finishing it today after re-writing part of it. Thank you for sending me the copies of the letters I wrote to you a while back. This letter is the product of a 'stream of consciousness' approach without following an outline of what I want to say.

I have not gone to the library to consult the journal about item 23 on p. 156 in the AMS Semicentennial Vol I (a paper by Osgood is #23). In relation to Osgood I found an item of interest in the book by

Caratheodory–<u>Funktionentheorie</u> vol. 1, published by Birkhauser, 1950... Section 193, starting on p. 187 is called Der Satz (Theorem) von Osgood. The specific theorem, which is more general than Osgood's result, is in italics on p. 188. If a sequence of meromorphic functions is convergent at every point of a region G, there exists an open subset S of G that is everywhere dense in G and such that the sequence is continuously convergent on S.

At the bottom of p. 188 Caratheordory asserts that Osgood stated this theorem for the case in which the functions of the sequence are holomorphic (regular analytic).

The concept of continuous convergence, originating with Caratheodory, I believe, is closely related to uniform convergence. It is defined at the top of p. 172 in C's book, and the relationship between continuous convergence and uniform convergence is given on p. 177 (in italics).

I met Caratheodory while I was at Caltech in 1936 or 1937.

Find in another book of mine a reference to "the theorem of Stieltjes-Osgood." The book is by Stanisław Saks and Antoni Zygmund, titled <u>Analytic Functions</u>, in an English translation from the Polish original by E.J. Scott. It is vol. 28 in the series Monografie Matematyczne, 1952. (By the way, the l with a slash, ł, has a special sound more or less like w.) I bought this book while attending the International Congress of Mathematicians in Amsterdam in 1954. The reference to Stieltjes-Osgood is at the top of p. 119. It asserts that if W is a family of holomorphic functions is almost uniformly bounded in a region G it is a normal family. I won't give the definition of normality, (a concept capitalized on by Paul Montel), but to be almost uniformly bounded on G means being uniformly bounded on each closed subset of G.

No bibliographic references are given by Caratheodroy or Saks-Zygmund. I haven't identified papers by Stieltjes or Osgood that are implied by the Stieltjes-Osgood label. I don't recall any reference by Osgood to normal families but my memory may be failing me.

Next—Duhamel and infinitesimals. See Osgood's item 30 in the AMS article about Osgood. The reference is Annals of Math, 2nd series, vol. 4, 1903. Osgood deals with the subject in his Introduction to the Calculus (Macmillan 1927. See pp. 301–304), which I studied at Harvard in my freshman year. Before getting into that book I also studied Osgood-Graustein's book on plane and solid analytic geometry. I can't say that I was smitten by Duhamel's theorem, but I came to appreciate its usefulness and logical significance in applications of integral calculus to physics and geometry. The applications are indicated in following sections. Infinitesimals are defined and discussed first in Osgood's 81–91 before defining differentials. When I came to write a text-book on calculus, I decided to avoid entirely the word infinitesimal, but to retain the reason for introducing what I ultimately called Duhamel's principle, so that the limits of certain sums would be rigorously know to be certain definite integrals. I did what I thought best in three successive editions of Calculus, by Sherwood and Taylor, Prentice Hall, 1st ed. 1942, 2nd ed. 1946, 3rd ed 1954. There were slight modifications in the exposition in the 2nd and 3rd ed. On p. 324 of the 3rd ed. I called the principle Osgood's form of Duhamel's principle. My books were very successful and used widely at the best universities. But not many other writers of calculus texts ventured to use Duhamel. I also wrote about Duhamel in my Advanced Calculus. Ginn & Co. 1955, pp. 515–517. There I used the notion of partition weighting.

I think Osgood used the word infinitesimal with quite a bit of care. He most certainly did not think of an infinitesimal as an infinitely small non-zero quantity. He inveighed against what he called "those horrible little zeros." The vague use of the word infinitesimal goes back to the time of Leibniz, when the derivative was conceived of as an ultimate ratio of vanishingly small quantities. The French, many of whom knew better, took the liberty of using the notion of un infiniment petit. It was Abraham Robinson whom I brought to UCLA in the early 1960's, who showed how to construct a legitimization of the intuition of Leibniz by creating an enlargement of the real number system in what Robinson called non-standard analysis, so that the derivative is an actual quotient of "ideal elements." See Robinson's Non-Standard Analysis 1966 North Holland Pub Co, especially Chapter III.

I also studied Osgood's Advanced Calculus (Macmillan 1929). It is a pot-pourri from which one gets exposure to a great many things.

Neither Osgood's Intro. to the Calc. nor his Advanced Calc. contains a systematic rigorous theory based on a fully explicit treatment of the real number system. The nearest Osgood comes, in the Intro. to the Calculus, to the completeness of the field of real numbers, is where he deals sparingly with the existence of a least upper bound of a nondecreasing sequence that is bounded above. He was willing to have students take for granted that a function that is continuous on a finite closed interval bounded in value, takes on its least upper bound and greatest lower bound, and assumes all values between those two bounds if they are different. All of these things are dealt with with perfect clarity and rigor in Chapter 1 of his Lehrbuch der Funktionentheorie. That is where I learned the basics of the theory of continuous functions of a real variable, when I took Osgood's famous course Math 13, Theory of Functions of a Complex Variable, in my 3rd or 4th year at Harvard. (I don't recall for sure which year). He lectured every detail of the course (in English, of course), but I could read German and I bought the book. (Vol 1, 5th ed. 1928).

When I came to write my Advanced Calculus I put all of the theory I had learned from Chapter 1 of the Lehrbuch into my book. But by that time I also knew about complete ordered fields and I thought the students ought to know what I knew. Most of the students in my year-course on Advanced Calculus were 3rd year students—some were seniors and a few were beginning grad. students.

I am sending you photocopies of pp. 192–201 from an unpublished MS autobiography. There you will find answers to the questions about Osgood that you asked. If you decide to use some of what I have written and mention your source, you should acknowledge that I gave you the information and say that I had great respect for Osgood as a teacher and mathematical scholar.

I am also sending you a copy of a letter written to me by Garrett Birkhoff in 1989. I came across it by chance as I was searching my correspondence files for a separate reason.

In my day at Harvard the Math Dep't gave a course—Math 4, I think—on statics and comparatively simple dynamics of particles and rigid bodies. One semester my teacher was Osgood; the other semester was by E.V. Huntington, as I recall. Then I took Osgood's course based on selections from his Advanced Calculus. (Margin note: My teachers in the 2nd year of calculus were Marston Morse and M.H. Stone.) I think it was in either my junior or senior year that I took a 1-semester course in analytical dynamics, taught by Osgood. As I learned from later contacts with Osgood while I was a graduate student at Caltech, he was writing his book, Mechanics, (Macmillan, 1937). In that course and Math 4 I learned much of what Osgood put into his Mechanics. I recall, in particular, the beautiful treatment of the motion of a gyroscope, a spinning top, and a bicycle. Also Euler's equations of motion.

I had Julian Lowell Coolidge as my tutor. He was the Master of Lowell House in its initial year of operation. I lived in Lowell House. Coolidge had me read parts of Bôcher's Higher Algebra, which I did not greatly enjoy. Coolidge wanted to do all the stuff about quadratic forms into geometry. I also read one of L.E. Dickson's books on the Theory of Equations, and a French analyst's book on the theory of real functions. I could read French easily. This book did not go beyond 19th century analysis. I never really appreciated modern algebra until I read Van der Warden and Birkhoff and Mac Lane. At Caltech, under E.T. Bell, I studied Galois Theory from a book by Dickson and never really liked his exposition. Only after I learned about Hilbert space and self-adjoint operators from von Neumann and M.H. Stone (at Caltech) did I <u>really</u> understand the business I had read in Bôcher's book about principal axes of symmetry and the eigenvalues of symmetric matrices.

I knew of the existence of Professors Widder and J.L. Walsh, but I never took any courses from them and I only got to know them personally much later on. I don't recall ever hearing about Lebesgue and his measure and integral until I studied Lebesgue integration from a book by Titchmarsh when I was a grad student at Cal Tech.

For my senior honor's thesis at Harvard I wrote about Fourier series and the induction of heat. I read a book by H.S. Carslaw (Macmillan 1921) on the Mathematical Theory of the Conduction of Heat in Solids. I also consulted a book in German (Weber-Riemann) on the partial differential equations of mathematical physics. My choice of this thesis topic came as a result of a course I took in the Physics Dep't, where I learned (in a heuristic manner only) about solutions of problems by the separation of variables and expansion in orthogonal functions.

I can't answer the question about accounting for why Osgood did not supervise the doctoral candidacy of very many students. As my MS shows, Osgood could be genial with undergraduate students. I never mixed with graduate students in my time at Harvard. There were about four of my classmates who, as with me, took highest honors, but I was never intimate with them: Ralph Boas, Harold Duw, Bert Arthur Writer, and one whose name escapes me. (Note:

transcriber is unsure of the last two names.) I got to know Boas later when we were both post-docs at Princeton/Inst. Adv. Study. I never went to see any of my teachers at "office hour", except my tutor, so I never knew of 'scuttle-but'—about who was or was not a good sponsor. My experience at UCLA, where I supervised 15 or 16 Ph.D. candidates in the 18 years after I became a full professor, revealed that some faculty members were far more aggressive than others in 'latching onto' grad students. Among my best students were W.G. Bade, David C. Lay, Seymour Goldberg, but there are several others that I classed as very good.

It could be that the trends in "hot fields," not among Osgood's interests, took most of the grad students in directions away from Osgood. He may have suggested possible topics to quite a few students, but few of them could get worthwhile results.

As you may know, Lebesgue's doctoral thesis was very controversial, and for reasons we aren't aware of, Osgood may have been reluctant to jump into the new fields that it opened up.

I think I'll wind this up now.

Sincerely,

/s/Angus E. Taylor

Afterthoughts: I sent Osgood a copy of the first addition of the Sherwood-Taylor Calculus. He wrote me his favorable opinion. That edition was reviewed extensively in Science by G. Bailey Price—an almost ecstatic review.

Then the following thoughts flowed freely on 2/16/97.

I did not, at Harvard, have a glimmer of what was then in the wind: general point-set theory, abstract point-set topology, abstract vector spaces, functional analysis, and a vastly changing situation affecting the theory of real functions of a real variable. In the years from 1933 to 1965 I went through the stages of learning how to present to graduate students and the most exceptional undergraduates the modern and penetrating way of preparing themselves to become masters, scholars and creative users of what I called, in my book (1965) <u>General Theory of Functions and Integration</u> (now available from Dover). The prime movers in the early stages of the big change toward abstraction in mathematical analysis were Henri Lebesgue, Maurice Fréchet, Felix Hausdorff and Fréderic Riesz. There were other important contributors, of course, notably S. Banach. At a later stage, in the early 1930's, M.H. Stone and John von Neumann made abstract Hilbert space an essential feature of modern analysis. Hilbert himself was a fundamental contributor in his student of integral equations, but he did not go abstract. Also, I don't see that Hilbert was greatly influenced by Lebesgue's ideas. It was F. Riesz who discovered the relationship between the sequence space l^2 and the class L^2.

/s/AET.

A.4 February 16, 1997[2]

Dear Diann Porter,

Along with my extended reply to your original query, I enclose three copies of things I have published in the last decade. They may or may not be things you and/or Professor Howard will wish to take time to read.

---

[2]Included were a Book Review of *Scenes from the history of real functions*, by Fydor A. Medvedev, which appeared in the *Bulletin of the AMS* in 1993; Taylor's article in the *Dictionary of Scientific Biography* about Maurice Fréchet; and Taylor's "A Further Look at Maurice Fréchet" which appeared in 1988 in the *Giornate di Storia della Matematica*, edited by Massimo Galuzzi.

I am glad that you kindled an interest in my thinking back to my Harvard years, 1930–33.

/s/Angus T.

A.5 May 9, 1997

Dear Diann Porter,

This letter is in response to your sending me the draft of your chapter VI about Osgood's teaching. But first a word about my personal history after my faculty career at UCLA. I was appointed Vice-President—Academic Affairs for the 9-campus UC system on Sept. 1 1965, as a consequence of my role as head of the Academic Senate of the whole system in the year 1964–65. I served as Vice President and subsequently University Provost of the system until I was selected to be Acting Chancellor and then Chancellor of the UC campus at Santa Cruz. That was a short service: Feb. 1, '76–Sept. 1, '77, to start when the current Chancellor couldn't make a go of it. I chose to retire a month before my 67th birthday. Since then I have been writing about the history of mathematics and various university-related matters as well as memoirs. I can't recall if I told you of my lengthy study of the work of Maurice Fréchet and the development of abstract point set topology and abstract analysis (3 articles in the Archive for History of Exact Sciences).

Now to your draft. I found it interesting and well written. Your use of information from me is entirely satisfactory.

Do you know anything about correspondence and personal papers left by Osgood? The scope of your dissertation would not extend to an analysis of Osgood's personal and family life, but I have always been curious about his life with his first (German) wife and what led him to marry the former wife of Marston Morse. I visited the Osgoods at their home in a Cambridge suburb in December 1937 when my wife and I went up from Princeton to visit my brother, a member of the Economics Faculty at Harvard.

I observe from what you told me at the outset of our correspondence that you are a candidate for the Doctor of Arts degree. How many institutions offer that degree? I recall my days at UCLA when the idea of that degree was being talked about. At that time the history of science as a distinct discipline was rather new, I think. My son Kenneth (born 1941) is an historian of science on the faculty at Norman, Oklahoma. His Ph.D. is from Harvard. His thesis research was in the area of geologic science, but he has a very broad intellectual horizon. My elder son Gordon (born '38) is in the field of English and American literature, with a Berkeley Ph.D. Both sons are graduates of Harvard College.

Please keep me posted about your progress. I hope you will write about Osgood's deepest work.

Sincerely,

/s/Angus Taylor

A.6   June 6, 1997

Dear Diann Porter,

Thank you for your letter of May 26. Congratulations on the job at Tucson. I have been there. One of the famous faculty members there became the great expert on dendochronology—the use of tree-rings in dating past climate.

Osgood's work is certainly worth historical study. He had a big influence on me. I had first seen him with a full beard. But when he came back after his stay in Nevada while getting a divorce, he was clean-shaven—and looked very different. Morse left Harvard to go to the Institute in Princeton, presumably to avoid being Osgood's colleague.

Two further items about Osgood: In a course I took from him about advanced theoretical dynamics he got confused in the explanation of something. I think it was about Euler's equations. He (Osgood) walked on the dais to where he could gaze out a window of Sever Hall, and stood there a long time, silent. Then he came back to address the class. I could see that he had tears of embarrassment in his eyes.

On p. 206, in section 11 (in vol II) of O's Lehrbuch der Funktionentheorie, there is a theorem about the possibility of analytic continuation of an analytic function of two complex variables. I was interested in the subject and its extension into abstract Banach spaces while I was a National Research Fellow at Princeton (1937–38). I could not understand Osgood's proof of the theorem and I wan't the only person with the difficulty. I published a paper about this theorem in Annals of Math vol. 40, 1939. Professor S. Bochner found a gap in my proof, and gave an entirely different proof. I never closed the gap. It had seemed ok to me at the time, and the referee didn't question the paper.

My doctoral thesis is published (in part) in the journal Annali della R. Scuola Normale Superiore di Pisa, Sec. II, vol. VI, 1937. Prof A.D. Michal, my prof. at Cal Tech, suggested that I send my paper there. It was Michal who interested me in M. Fréchet. I wonder if you know of the three papers I published in the 1980's about the mathematical corpus of Fréchet. They are in the Archive for History of the Exact Sciences.

Were you able to get in touch with Mrs. J.L. Walsh (Liz) in Nashville? A very interesting woman (Walsh's second wife). She rose to a high position in the military.

Enough of this from an aging man —

Sincerely,

/s/Angus Taylor

A.7 September 12, 1997[3]

Dr. Diann Porter

Dear Dr. Porter, or as I'll venture to address you—Diann. This will be a rambling letter. Thank you for giving me a copy of your dissertation. I've skimmed through it and read parts of it carefully. It is informative and well planned and executed.

...

I don't know anything about the relations between Osgood and Wiener. I didn't know Wiener in my Harvard days—1929–33, but I encountered him sometimes in later years. I never saw any sign of anti-Semitism in Osgood. But George David Birkhoff was openly anti-Semitic. I knew both Birkhoffs—father and son. Garrett was one year ahead of me. We were both in Lowell House at Harvard.

In my day at Harvard analytical mechanics was important in the curriculum. Math 4 was a year course in statics and the dynamics of a particle and the dynamics of rigid two-dimensional bodies subject to force vectors in the plane of the masses. Math 8 was

---

[3]This, the final letter in the correspondence, has been excerpted.

about general kinematics and dynamical systems. Euler's geometrical equations and Euler's dynamical equations were studied. Also Hamilton's principle, the principle of least motion and Lagrange's equation. There was also a course in potential theory, based on the book by O.D. Kellogg. (a beautiful book).

Osgood's book on Mechanics was not finished until after he retired from Harvard. For several years I taught potential theory and mechanics at UCLA, but it got crowded out as our doctoral program began to concentrate on pure mathematics more and more. My impression is that the more advanced Eulerian work in dynamics has almost disappeared from the curriculum in both mathematics and physics. But of course it must be learned somewhere, for travel to the moon and Mars.

. . .

I am a bit curious about the direction of your academic aspiration—as between teaching and research and as between intellectual history more broadly than for mathematics exclusively. My own bent is toward a rather broad liberal education. I think over-emphasis on research by faculty members is bad for the quality of life in our society. Undergraduate education deserves the attention of great thinkers to stimulate students.

I am nearly finished with a memoir about a part of my career—from 1938 when I went to UCLA as an Instructor at $2000/year. After a post doc year at Princeton, to early 1967, when I was the Academic Vice President of the 9-campus UC system and Clark Kerr, the President of UC, was fired by the Board of Regents, as part of the handiwork of Ronald Reagan. To keep the book within bounds, I've selected just this limited part of my career. When I finally retired, on Sept. 1, 1977, I was almost 66, after a short but eventful and enjoyable period as Chancellor at UC Santa Cruz (to handle a crisis there brought on by an inadequate Chancellor), I could have returned to Berkeley as Provost of the UC System, but I wanted some free years with my wife. She died while we were in France in 1982. Since then my life is made up in part by writing and in part by enjoying my three children and their families. (Loving detail about family omitted.)

...

I said this would be a rambling letter. It certainly is. Your thesis project stirred up my memories of Harvard and piqued my curiosity about your career. I've been in Tucson. UC's president Charles Hitch, a very close friend of mine who died a few years ago, of Alzheimer's, was a graduate of U. of A. I was down there a number of years ago for a November get-together with Hitch, his wife and friends.

Now you can be quit of me!

/s/Angus T.

www.ingramcontent.com/pod-product-compliance
Lightning Source LLC
Chambersburg PA
CBHW070557100426
42744CB00006B/322